SEARCHING FOR THE OLDEST STARS

SEARCHING

FOR THE

OLDEST STARS

ANCIENT RELICS FROM THE EARLY UNIVERSE

✦ ✦ ✦

ANNA FREBEL

TRANSLATED BY ANN M. HENTSCHEL

PRINCETON UNIVERSITY PRESS
PRINCETON AND OXFORD

Published by Princeton University Press, 41 William Street, Princeton,
New Jersey 08540

In the United Kingdom: Princeton University Press, 6 Oxford Street,
Woodstock, Oxfordshire OX20 1TW

press.princeton.edu
Jacket image courtesy of NASA/WMAP Science Team
All Rights Reserved

ISBN 978-0-691-16506-6

Library of Congress Control Number: 2015948026

British Library Cataloging-in-Publication Data is available

This book has been composed in Minion Pro and Trajan Pro
Printed on acid-free paper ∞

Printed in the United States of America

10 9 8 7 6 5 4 3 2 1

Searching for the oldest stars

Bernard of Chartres used to compare us to dwarfs perched on the shoulders of giants. He pointed out that we see more and farther than our predecessors, not because we have keener vision or greater height, but because we are lifted up and borne aloft on their gigantic stature.

—John of Salisbury, *The Metalogicon of John of Salisbury*

✦ ✦ ✦

In the same spirit, I dedicate my book to the women who lived and worked and changed the world before my time: to the women scientists as well as to my grandmothers and my mother.

CONTENTS ✦ ✦ ✦

PREFACE ✦ ✦ ✦

We regularly read about the latest scientific breakthroughs. We watch TV shows about science. We hope that our children will be good at math and science. However, all of this science stuff often seems rather abstract and "too difficult." Newspapers and magazines reporting on the latest findings usually concentrate on the best and most sellable results, and the results only. The methods and the people behind it are not described. This delivers a somewhat flawed view of science and scientists to the general public in a day and age when science and technology are the bedrock of our society. The public does not and cannot know what scientists do every day because their actual work remains invisible.

Indeed, what makes science hard to grasp is the lack of knowledge about the underlying motivation and inspiration for scientists' work. The thrill of scientific discovery cannot be shared without taking a closer look at how scientific results are obtained and what scientists do all day long. While technical and physical details are important for performing science, they are much less important for enjoying and comprehending it. After all, it is acceptable to enjoy a painting even if you cannot paint.

In high school I was an astronomy-loving teenager with just a basic math and physics background. I desired to become an astronomer, so I asked around for a scientist's job description. I wanted to experience astronomy and learn what astronomers do, both in daytime and at night. But nobody could answer my questions, and it is disappointing that this situation has not changed much since then. Thus my goal here is to answer these questions for those interested in astronomy, while engaging the reader and providing insights into the exciting field of stellar archaeology.

To reach the widest audience possible—from high school students to senior citizens—I opted to present a mixture of chapters of different levels and with varying, yet closely related, content. This approach clearly sets this book apart from other popular science books. For a person new to astronomy there are initial chapters about my journey and how my line of research arose. There are also observatory stories that humorously introduce what can go right and wrong when collecting data of celestial objects. Amateur astronomers will enjoy the in-depth chapters about various nucleosynthesis processes, spectroscopy, and the very first stars in the Universe. To keep everything flexible, each chapter can be read on its own, and chapters can be skipped without issue. Readers could even read the chapters backward and still get the story. However, the chapters do build on themselves pedagogically, such that curious novices can easily navigate the entire book, including the more detailed material described in chapters 5, 7, and 9.

My early working title for this book was "Paying Homage to the Stars," which expressed my feelings toward my work, fueled my writing, and helped me express my love for astronomy and observational research on paper. Today, I continue this quest of illustrating the beauty of astronomy for the public by filming video clips about our results and observations with the Magellan Telescopes in Chile. Offering an enhanced experience, they are an excellent complement to the book. They are available on my website (http://www.annafrebel.com) and on YouTube.

At the end of this cosmic journey I would like to thank Dr. Jörg Bong of S. Fischerverlag. His wonderful persistence eventually convinced me that I should write a book about stars in my native German. Dr. Alexander Roesler and the team of Fischerverlag then accompanied me on my journey toward authorship. Thank you all, especially for the enjoyable conversations in Frankfurt, New York, and Chile, which encouraged me to write. Furthermore, I thank my mother, Barbara Frebel, for repeatedly checking my chapters for inconsistencies. I thank my father, Horst Frebel, for providing additional assistance.

I thank Ingrid Gnerlich of Princeton University Press for helping to make the English version a reality, and Ann Hentschel for providing a first pass translation. Prof. Norbert Christlieb, Gregory Dooley, Dr. Heather Jacobson, Alexander Ji, Dr. Amanda Karakas, and Prof. John Norris generously provided comments that improved the manuscript. Finally, I am indebted to my infant son Philip for always being patient with his mom while she worked on the translation.

Regarding my passion for astronomy, I have always been actively supported by Dr. Martin Federspiel and Dr. Wolfgang Löffler. I am grateful not only for their comments on the (German) manuscript but also for sharing and constantly encouraging my love of stars. They accompanied me from the very beginning on my path into astronomy and eventually metal-poor stars. Last but not least, I thank my many wonderful colleagues, most of all Prof. John Norris and Prof. Norbert Christlieb, as well as Dr. Christopher Thom and all my students and postdocs for always making my research fun and enjoyable. It would not be the same without you!

Anna Frebel
Cambridge, Massachusetts
January 2015

AN INTRODUCTORY REMARK ✦ ✦ ✦

This is how my own journey began.

I have often been asked why stars and the Universe interest me so intensely. I cannot answer this better than explain why blue is my favorite color. It has simply always been that way.

Stars have fascinated me beyond words for as long as I can remember. At 14, I decided to become an astronomer, to learn more about stars, to discover where they come from and what is occurring in their interiors. Yet the path was still unclear to me. But my dream was to discover something new, something that exists beyond our Earth, out there in the Universe that had never been known before. I also wanted to find out what really makes the world go round.

This desire immensely motivated me, and so at age 15 I was overjoyed to intern with astronomers at the University of Basel. There, I learned directly from the scientists what sort of tasks constitute astrophysicists' daily routine. The experiments from the university's introductory astronomy class helped me learn many concepts and theoretical fundamentals about stars, galaxies, and cosmology. Equipped with this knowledge, at 17 I wrote a 55-page paper titled "Analysis of Color-Magnitude Diagrams of Selected Star Clusters from the Viewpoint of Stellar Evolution."

Even before my university physics studies had begun, I had come closer to fulfilling my dream of studying the stars and the Universe as an astronomer. Today I cross the globe several times each year to search for the oldest stars utilizing the world's largest telescopes.

SEARCHING FOR THE OLDEST STARS

CHAPTER 1

WHAT IS STELLAR ARCHAEOLOGY?

To understand the many details and the prevalent chemical and physical processes in the Universe, we will embark on a cosmic journey through space and time. It starts directly with the Big Bang and will lead us from there to the present. As can be seen in Figure 1.1, we will first acquaint ourselves with the cosmic origin of an apple and from there also with that of the chemical elements. The most ancient stars from the time shortly after the Big Bang will assist us on this journey. They demonstrate that we humans are all children of the cosmos. Made mostly of star dust, we even carry small amounts of Big Bang material inside ourselves.

The American astronomer Carl Sagan once said, "If you wish to make an apple pie from scratch, you must first invent the Universe." The elements composing an apple are in fact the result of a cosmic production process that lasted billions of years. Astronomers call this the chemical evolution of the Universe. The atoms of an apple were first generated by processes of nuclear fusion in the hot cores of stars eons ago. By baking an apple pie we change the order of the atoms inside the apples' molecules, but the atoms themselves remain unchanged. To change one kind of atom into another, our kitchens would need to be equipped with nuclear reactors.

The elements hydrogen and helium were formed in the very early phases of the Big Bang and provide the basic material structure of the Universe. Soon afterward, the cosmic cooking of the other elements began. This is how all the elements were ultimately generated to form the basis for the emergence and evolution of life, and hence also of human beings. For humans and organic matter in general, carbon plays a crucial role, so our existence depends on the stars that synthesized that carbon. As humans, we thus have surprisingly close ties to the evolutionary history of the chemical elements.

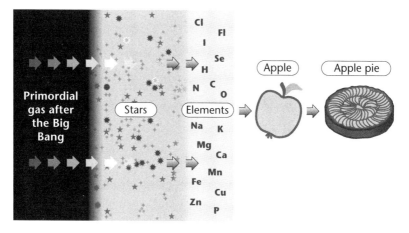

Figure 1.1. The cosmic origin of an apple. (*Source*: Peter Palm)

By analyzing the different chemical and physical processes involved in this evolution, astronomers can inch their way closer to understanding the nature of the whole Universe. Plate 1.A outlines this evolution. But let us start at the beginning of the story.

1.1 The First Minutes after the Big Bang

We often use concepts like space and time, temperature and density without considering whether there ever existed a "before" this space or a "before" this time. Our physical understanding of the Universe begins just tiny fractions of a second after the Big Bang, which should be considered the beginning of space and time. What really existed before and right at the beginning remains a mystery. "Big Bang" simply represents this indescribable initial state.

We do know, though, that immediately after the Big Bang the Universe was extremely hot and consisted of a thick soup of various kinds of tiny particles. During the minutes that followed, protons, neutrons, and electrons—the building blocks of atoms—formed. The Universe then expanded rapidly and quickly cooled in the process. The only chemical element existing up to that point had been hydrogen (atomic number 1). To be more precise, only hydrogen nuclei, that is, protons, existed. After two to three minutes the temperature had dropped to one billion degrees. The first nuclei heavier than hydrogen, including deuterium, were formed. Deuterium is also called "heavy hydrogen" because it has the same atomic number as hydrogen, but it is composed of one proton and one neutron.

From deuterium, the first helium nuclei (atomic number 2) formed, consisting of two protons and two neutrons. During the first two minutes, when the temperature had been even higher, helium had also been forming directly from four protons. But those helium nuclei were immediately destroyed by highly energetic gamma radiation. The detour via deuterium at the cooler temperature of about one billion degrees then finally led to the formation of larger quantities of helium.

The collisions of several helium nuclei caused the third heaviest element, lithium (atomic number 3), to occasionally form, albeit in only extremely small amounts. The Universe was then composed of three elements: hydrogen, helium, and lithium. Roughly 75% of the total mass consisted of hydrogen, 25% helium, and merely 0.000000002% lithium. For comparison, expressing this distribution in percentages of hydrogen and helium atoms, there would be 92% hydrogen atoms and just 8% helium atoms because helium is four times heavier than hydrogen. Lithium, in turn, constitutes just a minuscule fraction.

The first phase of element synthesis was complete just three minutes after the Big Bang. The Universe had cooled down too far for continued nuclear fusion with hydrogen and helium. But for life to later evolve in the Universe and for humans to emerge, these three chemical elements were not enough. The elements needed to sustain life, including carbon, nitrogen, oxygen, and iron, as well as all other elements in the periodic table, were still missing. Those were later built up, nucleus by nucleus, inside stars over billions of years. Only the interiors of stars are hot enough for heavier elements to be successively synthesized from the available lighter elements, such as hydrogen and helium, and others, as time went on.

These stars, and later also galaxies, had to emerge first, however. For that, the positive, electron-less, atomic nuclei had to combine with the free electrons whizzing about the Universe to form neutral atoms. For quite some time after the Big Bang, these atomic nuclei, free electrons, and also photons were racing about in a cosmic jumble. The energy and direction of the photons were constantly being diverted—scattered—by free electrons. Hence, this soup of particles and rays was fairly opaque, similar to water droplets in the pouring rain or thick fog.

About 380,000 years after the Big Bang, the Universe had grown so much while cooling down that a fundamental change occurred when it reached about 3,000 K. The nuclei and electrons were moving slowly enough by then that the positively charged nuclei could capture the negatively charged electrons to bind them permanently. The photons that had been flying around since the Big Bang suddenly had much less chance of being scattered. Consequently, matter and radiation separated and the opaque Universe became transparent for the first time.

At last, the photons were liberated from the labyrinth of electrons and could traverse long distances unhindered. The photons from the early Universe are still flying around today—referred to as cosmic background radiation. They constitute the faint residual glow of the Big Bang from almost 14 billion years ago—the last glimmer of a gigantic cosmic firework.

Since becoming transparent, the Universe has grown 1,100 times larger. The energy density of the cosmic background radiation decreased as the Universe's volume increased. For that reason the temperature of the background radiation reaching us today is not 3,000 K anymore but just 2.7 K. Since the Big Bang, the Universe has come fairly close to absolute zero, 0 K, or −455 °F. As it continues to expand, someday in the very far future it will reach absolute zero temperature.

The Universe's background radiation was actually discovered by chance in 1964 by the American radio astronomers Arno Penzias and Robert Wilson, although others had previously predicted its existence. The two scientists received the Nobel Prize in 1978 for their work. Another Nobel Prize was awarded to the American astrophysicists George Smoot and John Mather in 2006. Together with their team, they obtained the first precise measurements of cosmic background radiation using the space satellite COBE (Cosmic Microwave Background Explorer) and were able to determine its structure and extension in space. These and other measurements taken with the Wilkinson Microwave Anisotropy Probe (WMAP) satellite (shown in Plate 1.B) provide confirmation of the Universe having gone through an extremely hot phase when it was occupying an immensely small space—in other words, the

Big Bang. The team led by Smoot and Mather was able to prove the existence of a very slight clumping of matter 380,000 years after the Big Bang, the time when the cosmic background radiation originated. Those early lumps were the condensation seeds of all later cosmic structures, in particular those of galaxies.

A few hundred million years passed before the Universe completely changed its characteristics yet again. The "dark ages" that had persisted since the atomic nuclei had begun capturing electrons came to an end. The first stars in the Universe emerged from the giant and increasingly clumpy clouds of gas. They were composed of just the hydrogen, helium, and lithium of the primordial soup left behind after the Big Bang. This way the cosmos was lit up for the very first time. The UV light emitted by these stars led to the ionization of neutral atoms in the gas clouds. The intense stellar radiation had dislodged the electrons from their atoms. The very existence of the first stars had thus altered the conditions for the formation of subsequent stars. As a result, star formation continued more efficiently. Greater and greater numbers of stars formed, and, together with the gas, they arranged themselves in huge clouds of stars known as galaxies.

In their hot interiors, the first stars synthesized chemical elements heavier than hydrogen and helium. This production of additional elements led to significant changes in the Universe yet again. From that time on, countless stars began to chemically enrich the surrounding gas in their galaxies. After about nine billion years, enough of the elements had accumulated for the formation of our Sun along with its planets in a galaxy that we call the Milky Way. Our planet Earth was made from a substantial amount of iron and other elements that had to be first synthesized in stars.

At present, after 13.8 billion years of cosmic evolution, the mass fraction of the elements from lithium to uranium is roughly 2%. When the Sun was born about 4.6 billion years ago, it was about 1.5%. Apart from any lithium, all this material was produced in stars. For this reason, stars, especially the oldest ones, are the key to understanding exactly how the chemical diversity currently present in the cosmos developed over time.

The requirements for the existence of life were met by the time the Sun formed. A human being is mostly made of water, H_2O, composed of oxygen made inside stars and hydrogen from the Big Bang. An oxygen

A cosmic treat!

Figure 1.2. Big Bang soda—a sales hit of cosmic proportions! Ingredients: water, sugar, and citric acid, composed of hydrogen, carbon, and oxygen as well as some trace amounts of calcium, iron, magnesium, phosphorous, potassium, and zinc. Origin: Big Bang (hydrogen), red giant stars (carbon), and supernova explosions of massive stars (oxygen and heavier elements). (*Source*: Peter Palm)

atom is about 16 times heavier than a hydrogen atom, so in a water molecule the mass ratio of hydrogen to oxygen is 1 : 8. Since our body weight comprises 65% water, this means that 8% (i.e., one-twelfth) is hydrogen. Voilà. We ourselves are part of the Big Bang as the hydrogen inside us originated from within the first minutes of the Universe. A person weighing 75 kg is thus carrying around about 6 kg of Big Bang hydrogen. Babies' water content is even higher, almost 90%, so a baby weighing 3.5 kg contains 370 grams (11%) of Big Bang hydrogen, which roughly corresponds to the weight of a full can of soda. As illustrated in Figure 1.2, we consume these and other elements every time we drink, for instance, a tasty lemonade.

The chemical elements that constitute the molecules that compose our bodies are billions of years older than the few years that have elapsed since each of us was born. So how do astronomers explore this cosmic past?

1.2 Stellar Archaeology

In the same way that archaeologists search for relics of earlier civilizations and epochs, stellar archaeology explores the early cosmos by means of old stars. Of course it is not a matter of digging in the dust or dirt somewhere in a desert under the blazing sun but instead searching the night sky for stars dating to the time shortly after the Big Bang. The main requirement is a sky survey, which corresponds to the selection of an excavation site. Sky survey data list all objects observed with a dedicated telescope in a particular region of the sky, along with their positions, brightnesses, and other characteristics, such as color.

Then begins the laborious task of digging through all the entries in those huge star catalogs with the help of computer algorithms, the excavations so to speak. Figure 1.3 illustrates this approach. At some point the astronomer finds a potentially interesting object, which is set aside for a more detailed inspection in a subsequent step of the entire selection procedure. A small or medium-sized telescope with a mirror of 2 to 4 m in diameter is needed for this task.

Then the whole procedure has to be repeated. Most stars are not interesting enough for further observation. Only the best, most promising objects are observed subsequently, but then with one of the world's largest telescopes. Still, astronomers need a bit of luck as well. In the end, only few such objects turn out to be truly important contributors to the advancement of science. But this is precisely the goal, unearthing those ancient stars.

Comprehensive surveys of the Milky Way exist to provide astronomers with plenty of data and to help them reconstruct the long evolutionary history of the Universe almost all the way back to the beginning. Plate 1.C depicts the Andromeda galaxy, our slightly more massive sister galaxy, to show what the Milky Way might look like when viewed from far away. Every new finding about the structure and evolution of the Milky Way also leads to a more complete understanding of other galaxies such as Andromeda.

As stellar archaeologists, we primarily study the chemical composition of the oldest stars found in the Milky Way. This idea is illustrated in Figure 1.4. It means that our abundance measurements of the elements in those stars help us reconstruct how the chemical elements evolved

Figure 1.3. The "excavation" of ancient stars. Large sky catalogs are needed to locate some of those rare objects. (*Source*: Peter Palm)

throughout cosmic history, almost as far back as the Big Bang. This approach gives us a glimpse into our home galaxy's earliest epochs and lets us draw specific conclusions about how stars and even galaxies formed and evolved in the early Universe.

In this context, astronomers use old stars to answer a broad range of fundamental questions. This work resembles archaeologists excavating the remains of a Stone Age settlement to reconstruct how and in what environment these people used to live. Stellar archaeology does exactly the same thing. From observational data, it reconstructs the characteristic features of the first gigantic supernova explosions that expelled freshly synthesized elements into their surroundings like immense fountains. Which elements did they produce and in what quantities? Can the conditions for the emergence of the earliest stars and galaxies be deduced from these results?

We have not yet completely analyzed all of the sky surveys of the past decade to systematically track down ancient stars, and new projects look even more promising. The field of stellar archaeology is thus buzzing with excitement. The Australian SkyMapper telescope and other surveys are currently producing enormous amounts of data. Such large-scale observations of the Southern Hemisphere sky are bound to produce countless discoveries of ancient stars in the Milky Way's outer region, the so-called halo. New dwarf galaxies will likely be discovered and immense strung-out stellar streams, often taking up huge sections of the

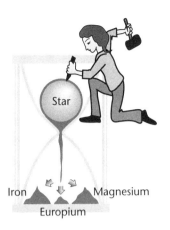

Figure 1.4. A stellar archaeologist's task: to determine the chemical composition of ancient stars. It calls for careful work and patience. (*Source*: Peter Palm)

sky, will be found. Plate 1.D shows the Field of Streams with various stellar streams found in the Northern Hemisphere that wrap around the Milky Way. Many of the recently discovered faint dwarf galaxies are labeled as well. All these data will help us explore even better the chemical and dynamic processes that led to the formation of stars and galaxies.

Another complementary approach to studying the early history of the Universe besides using ancient Galactic stars is observing extremely distant galaxies and gas clouds. This approach is widely used and is pretty well known to the public, partly because of the spectacular images of the farthest galaxies that, for example, the Hubble Space Telescope has been delivering since 1990. The Hubble Space Telescope can be seen in Plate 1.B, along with some of the most impressive images it produced.

Those extremely faraway gaseous objects emitted their light as young galaxies in the early phases of the cosmos. Since the speed of light is finite, it took billions of years for their light to finally reach us. This method offers a way to directly examine the past. We thus know that at least some stars already existed about 700 million years after the Big Bang. However, unlike stellar archaeology, this technique can provide only a limited amount of detailed knowledge about the chemical composition of the earliest stars formed and about the production of the chemical elements in their interiors.

Before we devote ourselves further to chemical evolution and the history of our Milky Way, let us first glance at the historical course of research about stars, their luminosities, and their element synthesis.

TWO CENTURIES OF PURSUING STARS

For thousands of years people have been looking at the celestial night sky to admire the myriad little lights. Each of those tiny sparkles is a star in our Milky Way. The immense brightness of these Galactic suns allows them to cast their light many light-years afar—as if they are our Galaxy's street lights. They appear peaceful and a bit mysterious on the night sky and give us an inkling of the vastness of the cosmos. Other distant galaxies have countless stars too. They light up their own home galaxies—like a football stadium alight in the distance at night. When we observe galaxies, the stars inside them thus guide us even farther beyond, into the seemingly infinite expanse of the Universe.

But how is it possible for stars to shine so powerfully and for as long as billions of years? What exactly is happening in the Sun's interior so that we can receive its light day after day? After all, sunshine is enormously important for us humans and for the entire evolution of life on Earth.

Surprisingly, it has only been about 75 years since the answers to these fundamental questions were finally found. What really happens inside the Sun, and therefore inside all other stars, has been known for only a very short time. As is often the case in science, the path to this new knowledge was marked by many discoveries that fell into place, piece by piece, like the stones of a large mosaic, built over the course of years, to finally yield the full picture. Looking back to this period, it is fascinating how the physical characteristics of stars were systematically explored, analyzed, confirmed, and sometimes even disproved. It must have been an exciting time in physics when so many now common scientific concepts were developed.

Let us now pretend to be flies on the walls of the offices, laboratories, and observatories of the physicists, mathematicians, and astronomers of the early 19th century and secretly watch them as they make their discoveries. This was the time when the first theoretical foundations were developed that paved the way to our understanding why the Sun shines each day.

2.1 First Glimpses of Stellar Rainbows

The long road to solving the puzzle of the Sun's energy source began with Isaac Newton of Britain, who discovered in 1666 that it is possible to separate the colors of sunlight, for instance, by passing it through a prism. The colors contained in light can be fanned out on a screen behind the prism into what is called a spectrum. In the case of sunlight the resulting spectral colors are red, orange, yellow, green, blue, and purple. A rainbow is a naturally occurring spectrum when raindrops act as the prism. Physically, the impressions of different colors in the eye correspond to specific wavelengths of light. Red light, for example, has a longer wavelength than blue light.

This work was continued by Joseph Fraunhofer at the beginning of the 19th century. The German optics expert developed various optical instruments, such as finely polished lenses, prisms, and telescopes, to carry out systematic spectroscopic studies of light. The young Fraunhofer experimented with different light sources, such as fire and in 1814 also with sunlight, to artificially produce particular colors. While doing so he noticed that the solar spectrum is "adorned" with countless dark lines of varying intensity. They seem to divide the color spectrum into many small segments, as if something had "stolen" the light at those wavelengths. Figure 2.1 illustrates such a spectrum. He began to meticulously catalog these vertical lines and their wavelengths. He designated the strongest lines with the letters A to K and labeled some other weaker lines with lowercase letters. This way he identified over 500 spectral lines. Thanks to improved instrumentation, today we know of thousands of these lines in the solar spectrum.

Other scientists had also noticed some dark stripes in the spectrum of sunlight prior to Fraunhofer. For example, the English chemist William Wollaston noted these already in 1802. At the time, though, his observations were dismissed as unimportant. Fraunhofer realized that

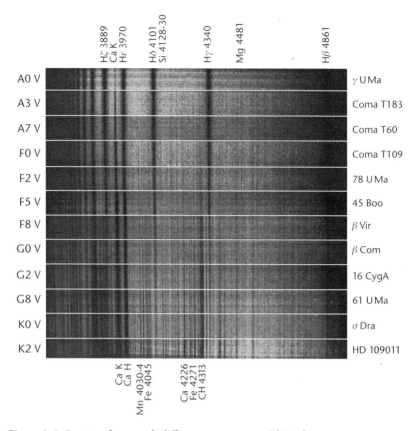

Figure 2.1. Spectra of stars with different temperatures. This is how spectra are recorded at the telescope. In 1814, Fraunhofer had already observed the many dark absorption lines. Astronomer Annie Jump Cannon classified spectra like these based on their apparent line strengths. Spectral classes (*left*) are described in sections 2.3 and 7.2. (*Source*: Peter Palm; reproduction of spectra from Abt et al., *An Atlas of Low-Dispersion Grating Stellar Spectra*, Tucson, AZ: Kitt Peak National Observatory, 1968)

the lines are a property of sunlight because he found exactly the same spectral signatures cast from the light of clouds, the Moon, and the planets. As these objects do not emit light on their own but merely reflect sunlight, a characteristic of sunlight had to be involved. But nobody knew how to explain the lines yet. Unbeknownst to Fraunhofer, those dark stripes, still known today as Fraunhofer lines, were one of the most important discoveries of science.

One can imagine these lines as similar to the bar code on a package of cookies. An astonishing amount of information is packed onto a very limited area that is decodable at checkout. Similarly a stellar spectrum allows astronomers to decipher vast amounts of coded information about a star when analyzing the light of its "spectral bar code." This is why spectroscopy is a major field of research in astronomy. Stellar archaeology and work on the chemical composition of ancient stars is also based on spectroscopic observations.

The breakthrough that ultimately led to an explanation of these observations happened only about 45 years after Fraunhofer labeled the spectral lines in the solar spectrum. Around 1853, a Swedish physicist, Anders Jonas Ångström, proposed different theories about the light emitted by gases and their corresponding spectra. Similar works on the spectral properties of the light of incandescent metals and gases were published shortly afterward by the American scientist David Alter. This new knowledge slowly caught on, and it took until 1859 for the physical causes of the Fraunhofer lines to gradually be revealed. Finally, the German physicist Gustav Kirchhoff and the chemist Robert Bunsen demonstrated in laboratory experiments that some Fraunhofer lines appeared at exactly the same wavelengths as the bright emission lines in the spectra of glowing metals.

It became obvious that the spectral lines of the substances analyzed in the lab had to be the same as the ones found in stellar spectra. Evidently, each substance has its own unmistakable pattern of spectral lines. This finding led to the emergence of the field of spectroscopy and to the discovery of new elements such as cesium (in 1860) and rubidium (in 1861). In the end, Kirchhoff and Bunsen were able to deduce that the dark lines in the spectrum of the Sun are attributable to the absorption of light by the chemical elements present in the solar atmosphere. They had found the "fingerprints," so to speak, of atoms and thus of individual elements. This was a magnificent scientific breakthrough that forever closely linked physics, chemistry, and astronomy. Chemical analysis of different objects on Earth as well as in space suddenly became possible. Consequently, Kirchhoff became the first to compare the solar spectrum in detail with the spectra of some 30 elements known at the time. He found that the Sun consisted of at least sodium, calcium, magnesium, chromium, iron, and nickel.

Kirchhoff continued to conduct fundamental research in spectroscopy in collaboration with Bunsen. Among other things, he combined the knowledge gained by Ångström and Alter from around 1855 concerning the radiation of hot bodies and gases and their emissivity with his own discoveries and explanations of spectral absorption. The result was rules that are still valid today. They define in which cases a continuous spectrum is generated and when a spectrum should have emission or absorption lines.

In 1863, the Italian priest and astronomer Angelo Secchi began to systematically record and analyze stellar spectra. He wondered whether different stars have different compositions. This research extended Fraunhofer's work on the solar spectrum to include more distant stars. In total he analyzed some 4,000 spectra and found that all of them can be divided into particular groups and subgroups, based on how many absorption lines they have and how strong they are. In other words, a spectrum can be categorized by its morphological characteristics. He specifically found five groups of spectra that occurred very frequently. These five so-called Secchi classes became the first classification system for stellar spectra.

He noticed, among other things, that in particular molecular carbon causes broad absorption bands to appear in stellar spectra. For these special types of stars he introduced the class of "carbon stars," which is still used today. His entire classification scheme continues to play an important role in astronomy.

This crucial research led Secchi to become the first astronomer to provide proof that the Sun is, in fact, just like any other star. Another early and significant spectroscopic survey was conducted by the wealthy Englishman William Huggins and his wife, Margaret. They were both interested in astronomy and began spectroscopically to examine many stars, nebulae, and galaxies around 1860 using their private telescope in London. They were the first to realize that different objects in the cosmos exhibit different spectra. The spectra of some nebulae resembled the emission spectra of gases, while the spectra of galaxies looked more like those of stars. Based on their experience with stellar spectra, they concluded that although the spectra of stars often differ, all of them are composed of the same elements, namely, the elements that make up the

Sun and Earth. "Heaven" and "Earth" were thus made from the same material. This contradicted a doctrine by Aristotle that had been upheld for almost 2,000 years: everything "higher than the Moon" had been thought to be made of ether.

Around the same time, another Englishman, Norman Lockyer, became increasingly fascinated with spectroscopy. He too was able to study cosmic objects and their compositions using his own small telescope with an aperture of just 16 cm. In 1868, he noticed, just as the Frenchman Pierre Janssen had, a hitherto unknown, unidentified, relatively strong line in the spectrum of the Sun's corona. The line is located very close to Fraunhofer's sodium-D lines in the yellow region of the spectrum. Lockyer proposed that this "yellow" line be attributed to an as yet unknown element in the Sun. He named the element "helium" after the Greek word for sun (*helios*). What is the second lightest of all the elements was detected on Earth only some 10 years later. It is a great example of how stellar spectroscopy led to the discovery of a new element.

Then, in 1885, the Swiss mathematician Johann Balmer discovered that the four strong absorption lines of hydrogen—the lightest element—in the visible light form an interrelated series with weaker lines in the near-ultraviolet region. Their wavelengths could be described by a simple mathematical formula. In 1888, the Swede Johannes Rydberg independently developed a general mathematical formula that could also be applied to other line series of hydrogen located in the ultraviolet and infrared regions. The Balmer series of hydrogen is still called that today, and the Rydberg formula provides a simple way to calculate the wavelengths of hydrogen lines and those of other elements in stellar spectra. In general, the hydrogen lines are the most prominent spectral lines. For this reason these new calculations contributed substantially toward a comprehensive interpretation of spectra.

The collection of enormous amounts of new spectroscopic data of stars and other celestial objects pushed astronomy, and science in general, a major step forward. It even advanced the way the world was viewed at the time. Spectroscopy made it possible to study those foreign, faraway objects from beyond Earth and to discover what they are made of. The analysis of light seemed to effortlessly bridge those immense distances. Suddenly, it became possible to pluck the stars from the sky.

2.2 Decoding Starlight

At the end of the 19th century people were readily using spectral lines for purposes of analysis and classification. How they formed remained a puzzle, though. New questions kept arising with regard to the many observed phenomena. Why did each chemical element have its own characteristic pattern of spectral lines? Why did some lines appear to be sharply focused while others looked diffuse? To solve these questions various scientists set out to investigate the inherent nature of the elements. The outcome was many novel concepts that ultimately led to the development of quantum mechanics.

The desire to explain the nature of the rather large stars—now possible thanks to spectroscopy—guided many contemporary scientists to look in the opposite direction, to the tiny building blocks of the elements, atoms. The time seemed ripe to consider how exactly the world, everything, was composed. Soon the theorists took center stage, assuming the long-held places of earlier experimenters such as Fraunhofer. They used their minds, and pencil and paper to probe the microcosmos opening before them. This had effects on the exploration and understanding of the macrocosmos, and hence astronomy.

As early as 1890, the German physicist Max Planck had been working on general radiation properties and what is referred to as a "black body." He found that such an idealized body emits a characteristic energy distribution that is reflected in its spectrum. The energy distribution of the radiation given off by a black body with a temperature of several thousand degrees actually resembles that of a star's energy distribution. The black body radiation has a maximum output that depends on its temperature. For a black body heated to about 6,000 K, this maximum lies in the green spectral region, which is where the Sun radiates most and where the human eye is most sensitive. In 1900, to describe this radiation, Planck put forward the extraordinary hypothesis that in any interaction between radiation and matter, energy can be exchanged only in discrete "portions." He called these portions "quanta." He postulated that each light quantum (a photon) has a specific energy proportional to the frequency of the light. For instance, high-energy quanta have a high frequency and consequently a short wavelength.

Following Planck's work, Albert Einstein further developed his ideas to show that electromagnetic waves can also be described as particles possessing specific energy quanta. By describing light as a particle and not as a wave, Einstein showed in 1905 that the new theory agreed with existing experimental data on the "photoelectric effect." Some materials emit electrons, but only when the material has been irradiated with a specific minimum energy that depends on the material and its properties. Those emitted electrons are called photoelectrons.

Einstein's explanation was as follows: A photoelectron has to absorb a specific minimum amount of energy to be freed. Only then can it leave the atom it is bound to. The individual light particles of the incident radiation need to transport this minimum energy: an electron absorbs one photon and its energy is transferred to it. Part of the absorbed energy, the minimum energy, is used to dislodge the electron from the atom. Any residual energy is converted into kinetic energy. Low-energy radiation of larger wavelengths than the limiting wavelength characteristic of the material cannot cause the release of any photoelectrons. The electrons would not receive sufficient energy to leave their atom. The energy of photoelectrons thus depends on just the energy of the incident radiation and not on its intensity. This quantization of energy into tiny portions, as found by Planck and Einstein, contradicted the previous notion that energy could be divided into portions of any amount.

This explanation led to an entirely new way of describing phenomena on subatomic scales. Einstein received the Nobel Prize in physics in 1921 for his work on the photoelectric effect. Today applications of the photoelectric effect are rather common, including in solar cells and digital camera sensors. To Einstein and his contemporaries the photoelectric effect and its implications were completely new and revolutionary. Light did not behave just as a wave, as had been assumed in experiments up to that point in time. Around 1861, the Scottish physicist James Clerk Maxwell had also described light as a wave with his Maxwell equations. But the photoelectric effect was best understood assuming that light *particles* transferred the energy necessary to dislodge an electron. This is exactly why in certain experiments light behaves like a wave while in others it appears to be a particle. This view, completely contradictory to everyday experience, is referred to as the wave-particle duality. In 1924, the French physicist Louis-Victor de Broglie showed that the wave-particle duality applies to every kind of matter. For instance, under certain conditions even electrons can act as waves.

The year 1905 was important for physics in many aspects. During his "Wunderjahr" ("miraculous year"), Einstein published four enormously important papers. Each of them changed our physical understanding of the world and gave science a tremendous boost. The explanation of the photoelectric effect was only the beginning. Next came his explanation of the Brownian motion of atoms and molecules in fluids.

Based on these works, he was able to provide the first indirect proof of the existence of atoms since the exact nature of atoms had remained unknown around 1900. Einstein also continued to study various processes of classical mechanics as they occur near or at the speed of light. This paper represented what came to be known as the special theory of relativity. It illustrated the various results of experimental attempts to verify the existence of the ether. For the first time, the speed of light was postulated to be a constant and finite quantity.

Finally, Einstein formulated the equivalence of mass and energy. It describes very generally that the mass of a physical body is also a measure of its energy. He showed that every particle has a "rest energy" in addition to its kinetic and potential energies. Therefore, massless particles should not have any "rest energy." Einstein's famous formula $E = mc^2$ was derived from the special theory of relativity as it applies to matter at rest. E stands for the internal energy of a body at rest, which equates with the product of the rest mass m times the square of the speed of light in vacuum c. In this equation, the speed of light is merely a factor to make the physical unit of mass agree with that of energy. The relativistic form is somewhat more complicated because the relativistic masses and energies have to be taken into account.

Max Planck had predicted long beforehand that a bound system would have less mass than the sum of its separate parts after the binding energy had been expended. Planck had likely been thinking of chemical reactions, but their binding energy was too low to be measurable. In those days, chemical reactions were often taken as a model for physical processes that were even less well known. Then, Einstein suggested that radioactive material, such as radium, might offer a way to test his theory. Radium, however, does not radiate strongly enough for such an experiment to succeed. Around 1905 no means were available to experimentally confirm Einstein's postulate. Only with the discovery of the positron antiparticle in 1932 could it be shown that the entire mass of a matter-antimatter pair is completely convertible into energy. Knowing about the equivalence of mass and energy was also of fundamental significance to the problem of why stars shine.

In 1913, the Danish physicist Niels Bohr introduced a new, sophisticated model to describe an atom and its structure. Following Einstein's 1905 papers outlining that atoms did actually exist, this was the next important step in unraveling the nature of atoms. Given the quantum theory of light, Bohr primarily wanted to understand the hydrogen atom and its characteristics. He developed a model for this simplest of all atoms and later extended it to incorporate heavier atoms in a shell model. He combined the atomic model proposed around 1911 by the British physicist Ernest Rutherford with Planck's concept of light quanta. Rutherford had been the first to assume that an atom is composed of a positively charged nucleus that is orbited by a corresponding number of negatively charged electrons to overall appear neutral. The electron and its negative charge had been known about since 1897, but the problem was that his classical description led to unstable atoms.

This new model of the atom, often called the Rutherford-Bohr model, still effectively describes atoms with just one electron, such as hydrogen, albeit in a simplified way. The grand accomplishment was the theoretical explanation of the Rydberg formula in describing the wavelengths of hydrogen lines in spectra. Bohr's suggestion was that negatively charged electrons revolve around the positively charged atomic nucleus like planets orbiting the Sun. An electrostatic force, rather than gravitation, would hold the system together. The configuration of such an atom is depicted in Figure 2.2. Using this quite comprehensive model, many of the already known characteristics of atoms could be explained for the first time. Accordingly, for heavier atoms with larger positively charged nuclei, many more electrons have to be present to keep the atom's overall charge neutral. Bohr's novel idea was that each electron orbit or "shell" could accommodate only a finite number of electrons. If a shell is completely full of electrons, another shell would have to be filled up next. Ultimately, the shell model explained many properties of the elements of the periodic table, including, for example, the increasing size of atoms

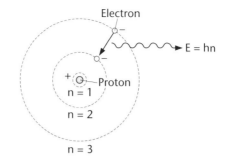

Figure 2.2. The Rutherford-Bohr model of the hydrogen atom (with atomic number $Z = 1$) or of an atom similar to hydrogen with $Z > 1$, where the negatively charged electron revolves around the positively charged nucleus along a given orbit. Electrons can jump back and forth between the different orbits and absorb or release quantized amounts of energy in the process. The orbital paths that the electrons may fly along are drawn as dashed lines. The atomic radius increases by n times 2, where n is the number of shells, counting from the inside out (the main quantum number in quantum mechanics). The transition from the third to the second shell is illustrated here. It corresponds to the first line of the so-called Balmer series of hydrogen that appears at 656 nm in the red spectral region. In this example, because one photon is being absorbed, an absorption line will appear in the stellar spectrum. (*Source*: Peter Palm)

in the periodic table when read from left to right, and the chemical inertness of the noble gases with their completely occupied outer shells. From then on, it was known that the number of electrons in the outer shells of atoms determined the spectral and chemical properties of the relevant element.

Bohr restricted the motion of the electrons to specific orbits that maintain finite distances from the nucleus and correspond to specific energies. The electrons on these orbits would fly around the nucleus without any loss of energy. Those orbits are today referred to as energy levels. Furthermore, it was postulated that electrons can jump back and forth between different orbits: When a jump from a higher energy orbit to one of lower energy occurs, a very specific amount of energy is released. Such emitted "light particles" have been called "photons" since then. If an electron finds its way from an orbit of lower energy to a higher energy one, energy has to be supplied—it "swallows" a photon of suitable energy. The photon's energy is therefore defined by the difference in energy between the two orbits, and its corresponding wavelength is described by the Rydberg formula.

Photons are elementary particles. They transfer electromagnetic force because they carry energy and momentum. They have no mass and thus propagate at the speed of light in any direction. For example, sunlight is relayed from the Sun to Earth by photons in about eight minutes. Light from the Sun needs that long to reach Earth.

By means of his atomic model, Bohr predicted in 1923, together with other scientists, the existence of the heavy element, hafnium (atomic number $Z = 72$). Not long afterward, it was indeed confirmed by experiments and named after the Latin word for Copenhagen, Hafnia. Hafnium is also measurable in some ancient stars. We shall come across it again later.

Planck's atomic model and quantization of light inspired the German physicist Werner Heisenberg in 1925 to propose an extension of classical mechanics that described the behavior of objects in the subatomic regime more precisely. Quantum mechanics was born. A second formulation of quantum mechanics based on the wave theory of light was developed independently in 1926 by the Austrian physicist Erwin Schrödinger. Schrödinger did not believe in the particle theory of light and proposed his famous "Schrödinger equation"—a wave equation—instead.

Quantum mechanics rapidly became the standard description of atomic physics. Its physical deductions and interpretations were developed further after Heisenberg discovered the uncertainty principle and Bohr introduced the principle of complementarity. In 1930, the English physicist Paul Dirac, in his *Principles of Quantum Mechanics*, even extended quantum mechanics to include the special theory of relativity. As a result, quantum mechanics became a comprehensive mathematical description of the particlelike and wavelike behavior of light and the interaction between matter and energy.

These findings and new characterization of atomic physics were rewarded with a number of Nobel Prizes, which demonstrates how fundamental these new results were. Taken together, they provided a new worldview of physics that would also pave the way toward resolving the problem of what the energy source of stars is.

The new theory was advantageous not just for understanding the cosmos. Today, we all frequently use quantum mechanics through technologies such as USB memory sticks, magnetic resonance imaging, transistors in computers, lasers, and electron microscopes. And we should not forget that modern digital cameras and the large photon detectors of current telescopes could not have been developed without knowledge of quantum mechanics. My research would have been considerably more tedious if I still had to use photographic plates for my observations, as all astronomers were doing right up to about 1990, rather than electronic devices. However, strictly speaking, even that old technique relies on quantum mechanics for a complete understanding of the processes taking place on a silver-bromide photographic plate.

Toward the end of the 1930s, the interconnectedness between different scientists and their fields of research were not only very obvious but also essential for progress. Some worked on small scales, on atoms, others on large scales, on stars. Some worked theoretically, others experimentally. All their findings built on one another, irrespective of whether they originated from chemistry, physics, or astronomy.

The period around 1900 is surely one of the best examples that there is no well-defined, obvious path to success in science. But the results from that time show, above all, that the scientists were very well connected to readily exchange and discuss their findings with each other. New knowledge spread rapidly, only to be immediately taken up by others and to be further "processed." This remains one of the basic principles of doing science today. Over the Internet, I collaborate regularly with colleagues from Australia, Europe, North America, and Japan.

As the secrets of light and photons had been revealed next came the study of the cosmos and its objects. The doors to outer space suddenly swung wide open, leading to yet another fundamental broadening of our worldview.

2.3 A New Perspective of the Cosmos

At the same time when quantum mechanics was at the forefront of research in Germany and Europe, astronomers in America were keenly investigating the Universe. Under Edward Charles Pickering's strict regime at the Harvard College Observatory, Henrietta Leavitt began work as one of several female research assistants in 1893. Their tasks consisted of performing various kinds of calculations and measurements of sky observations recorded on photographic plates. Because of their tedious work these women were referred to as "computers," but many discoveries showed they were capable of much more than just executing calculations.

Leavitt's predecessors included Williamina Fleming and Antonia Maury, who had worked for Pickering on extending Secchi's classification of stars. By around 1880, Pickering had become so dissatisfied with his male assistants that he declared even his housekeeper Fleming could do a better job. In fact he did hire her as a research assistant, and she did not disappoint him. Out of respect, Fleming completed every single task assigned to her—no matter how much work, at any time of the day. From then on, Pickering continued to hire more women, especially those who held diplomas in physics or astronomy. His ambition was the efficient development of an extensive system for the classification of stars. At that time, scientifically trained female staff members such as Maury and Leavitt were extremely cheap to employ and willing to work longer and harder than their male counterparts. In his somewhat peculiar way, Pickering thus demonstrated that women, too, could produce excellent scientific results.

One of Leavitt's duties was to catalog the brightness of stars. She found thousands of stars with variable luminosities in the Magellanic Clouds. The brighter objects among them seemed to have the longest periods. Conducting further observations, in 1912 she was able to confirm this behavior, known since that time as the period-luminosity relationship. A very special consequence soon emerged from this discovery. The intrinsic luminosity of those variable stars could now be calculated in addition to observations of their apparent brightness. Thus, with known distances of just a few such objects, the entire period-luminosity relation could be calibrated. Principally, this would enable distance determinations of other variable stars and change the course of astronomical research.

Just one year later, the Danish astronomer Ejnar Hertzsprung calibrated this relation using variable stars in the Milky Way. The foundations had been laid for measuring Galactic and extragalactic distances. At the time it was not yet known whether the Universe was any larger than the Milky Way itself. This method provided the opportunity to check whether anything existed beyond our Galaxy.

The newly found importance of variable stars led many astronomers, like the American astronomer Edwin Hubble, to conduct nightly observations of these pulsating stars. Hubble was looking for these objects in different diffuse nebulae, using the 2.5-m Hooker telescope at the top of Mount Wilson in Southern California. (Incidentally, right up until the mid-1960s, telescope observations were done exclusively by men, so that they could devote themselves to their work, without any distraction.) By applying the period-luminosity relation for variable stars, Hubble succeeded around 1923 in showing that those nebulae were far too distant to be part of the Milky Way system. They turned out to be individual galaxies far beyond the Milky Way.

Once again, new findings had transformed the contemporary worldview, this time by literally extending it—to unimaginable distances. Up to then, cosmic objects were assumed to have distances of little more than 100 light-years. The discovery of other galaxies hinted at the prospect of finding objects many millions of light-years away. This way, astronomers eventually found out that the Milky Way is just a relatively small object in a much larger Universe. Henceforth, the Universe was considered to be composed of thousands or even more galaxies, not just one. This discovery was initially met with some resistance by leading astronomers of the time, particularly Harlow Shapley of Harvard University. Nevertheless, the new finding held its own, and as a consequence the Andromeda nebula soon turned into the Andromeda galaxy, which we have come to know as our sister galaxy.

Leavitt died of cancer in 1921 at just 53 years of age. During her lifetime, this "computer" had received little acknowledgment for her fundamental research that significantly changed our understanding of the Universe. At an hourly wage of 25 to 30 cents, she earned a meager salary, less than a secretary. As others had not learned of her death, her name almost appeared on the list of Nobel Prize candidates for 1924. But that was too late for Leavitt. Since the prize is awarded only to living persons, she was not nominated.

Another "computer" was working on a new way to classify stellar spectra during the same period. The American astronomer Annie Jump Cannon started working for Pickering in 1896 after graduating with a degree in physics and astronomy. Her task as an astronomer was to catalog an extensive compilation of stellar spectra, called the Draper Catalogue, and to develop a classification system.

Cannon was the first to sort spectra by stellar temperature after realizing that a dependence existed between temperature and spectral line strength. Her new system later became famous as the Harvard classification scheme. She grouped the spectra into classes that are still in use today: O, B, A, F, G, K, and M—they are easily remembered by the mnemonic "Oh Be a Fine Girl/Guy, Kiss Me." These classes are based on hydrogen-line strength, one of the most important characteristics in stellar absorption spectra. O-type stars are the hottest (with surface temperatures of up to 50,000 K), whereas M-types are the coolest stars (sometimes only 2,000 K at the surface). Figure 2.3 illustrates example spectra for the various spectral classes in the form we currently use.

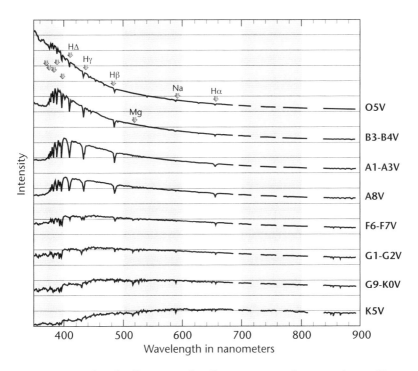

Figure 2.3. Examples of stellar spectra that illustrate present-day spectral types. Various absorption lines are indicated. Only specific lines are detectable in a given spectrum, depending on the surface temperature of the star. These one-dimensional spectra are obtained after data reduction and processing of the two-dimensional raw spectra (e.g., those seen in Figure 2.1). They are used for analysis. (*Source*: Peter Palm; reproduction of spectra from Silva and Cornell, *Astrophysical Journal 2*, supplemental series (1992): 865–881)

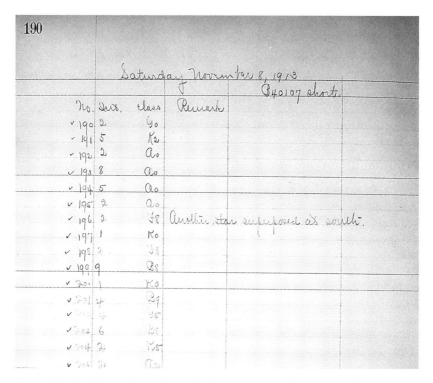

Figure 2.4. Annie Jump Cannon's notebook entries on star classifications from Saturday, November 8, 1913. Her notebooks are available in the Plate Stacks Archive (a collection of astronomical photographic plates) at the Harvard College Observatory in Cambridge, Massachusetts, USA. (*Source*: Anna Frebel; reproduction of one page from Annie Jump Cannon's notebooks, Harvard College Observatory Astronomical Plate Stacks Archive)

The huge collection of Cannon's bound laboratory notebooks containing her star classifications, as well as original photographic plates of the spectra intended for classification, can still be inspected at the Harvard College Observatory in Cambridge, Massachusetts. I held some of those notebooks, marked with the initials "AJC," in my own hands. One was dated November 8, 1913, and many pages record work done on a "Saturday." Sunday was the only day off. A sample of those entries is reproduced in Figure 2.4.

Cannon soon became known around the world for her classifications of over 200,000 stars. Her catalogs became standard reference works, and many important prizes and distinctions were awarded to her—

another a clear sign that women can do outstanding science just as well as men. One was an honorary doctorate from Oxford University in 1925. She was the first ever woman to be conferred with that title. Then, in 1931, the US National Academy of Sciences awarded her the prestigious Henry Draper Medal. Even Harvard acknowledged her achievements as worthy of recognition and in 1938 appointed her, at the age of 75, to the academic post of William Cranch Bond Astronomer, with the same rank as a professor. The American Astronomical Society had already introduced the Annie J. Cannon Award in 1934 to honor distinguished female astronomers in Cannon's name. Antonia Maury received it in 1943. Cannon published her huge catalog along with its various extended editions between 1901 and 1937. This work continued even after her death in 1941.

On the other side of the Atlantic, it was considerably more difficult for women to find scientific employment in astronomy. As a young woman in England, Cecilia Payne-Gaposchkin, for instance, was neither granted an actual university degree, despite finishing all the required scientific studies, nor allowed to work as a professional astronomer. At the advice of Harlow Shapley she immigrated to America to complete her doctoral thesis at Harvard College Observatory in 1925, under Shapley's guidance. She was only the second woman to do so. Shapley had assumed directorship of the observatory in 1921 and was more amenable to the idea of having professional women in science than was his predecessor Pickering.

In her thesis, Payne-Gaposchkin demonstrated that the variation in absorption line strengths in stellar spectra is attributable to the amount of ionized gas present. In addition, she found that the degree of ionization depends mainly on the gas temperature of the star. The line strength was therefore a measure of the surface temperature and not, as previously assumed, a measure of the stellar chemical abundances. This realization came roughly at the same time as the introduction of Cannon's spectral classifications. Consequently, from then on, it was possible to quantitatively ascribe spectral types to different surface temperatures. Payne-Gaposchkin concluded that stars are mainly composed of hydrogen and helium and do not have the same chemical composition as Earth, as was previously widely assumed. Indeed, heavier elements such as iron and silicon proved to be considerably rarer: she calculated that the number of atoms of heavier elements in a star is one million times less than that of hydrogen atoms. What a magnificent discovery!

Today, we know that a star does in fact comprise about 71% hydrogen, approximately 27% helium, and less than 2% heavier elements (by mass). This was one of the first things I learned about stars during my high school days. The American astronomer Otto Struve was right to describe Payne-Gaposchkin's dissertation as "undoubtedly the most brilliant PhD thesis ever written in astronomy." Struve was the founding director of the McDonald Observatory in West Texas from 1939 until 1950. I worked for the observatory, too, from 2006 to 2008, and Struve's statement still comes up in hallway conversations.

Payne-Gaposchkin's accomplishments paved the way for other women to enter astronomy. For her research she received the 1934 Annie J. Cannon Award, its first recipient. This prize is still being awarded to one young female astronomer every year. In 2010 I was the 41st recipient for my analyses of the chemical composition of old stars containing tiny traces of elements heavier than hydrogen and helium, and for my related research on the early Universe.

Interestingly, just like Maury, Cannon, and Payne-Gaposchkin, I was also employed at their workplace, from 2009 until the end of 2011, now known as the Harvard-Smithsonian Center for Astrophysics. Although my calculations were done electronically, there are reminders of the historic times: the portraits of many great female Harvard astronomers are hanging in the hallway of the old observatory building, telling the story of their outstanding accomplishments over the past century.

As the spectral classification of stars was being established by Cannon and Payne-Gaposchkin, Hubble in California was observing distant nebulae and galaxies. He measured their velocities with the help of spectroscopy. Rapidly moving objects show a shift in all their spectral lines compared to the wavelengths of a static object. In astronomy this shift is also called redshift because the spectral lines of galaxies are shifted to longer wavelengths, that is, toward the redder part of the spectrum. In 1929, Hubble discovered that almost all galaxies are moving away from Earth. Moreover, he noticed that the farther away a galaxy is from us, the more rapidly it is receding. These observations suggested that the Universe is expanding.

As we shall see later on, this discovery came at the right time when many other scientists were starting to work in cosmology, the study of the origin and evolution of the Universe. The velocity-distance relation has been obtained many times, with the goal to measure its proportionality constant, the Hubble constant. This is an empirical quantity that defines the rate of expansion, and therefore the age of the cosmos. Over the decades, measurements of the Hubble constant have triggered major differences of opinion and numerous and long-lasting controversies. Today, various methods of measuring it yield similar results that are neither as low nor as high as those contentious initial values.

By 1930, much progress had been made in astronomy. Fundamental characteristics of stars had been discovered along with the existence of other galaxies, both of which would prove to be essential contributions toward cosmology and our physical understanding of how energy is generated in stars.

2.4 Looking into the Hearts of Stars

Despite all the advancements made in physics and astronomy, the energy source of stars remained a mystery for the longest time. What process was responsible for sustaining the Sun's luminosity for billions of years? By 1907, chemical processes and gravitational contraction had been excluded, on the basis of theoretical works published by the Swiss physicist and astrophysicist Robert Emden. The discovery of radioactivity in 1896 heralded the era of nuclear physics, which would eventually solve the puzzle. But before that, another 20 years of intense research and numerous discoveries had to pass.

In 1896, the French physicist Henri Becquerel found that different uranium salts emit invisible radiation that could be detected after the exposure of a photographic plate. The Franco-Polish chemist and physicist Marie Curie and her husband Pierre Curie searched for the source of this emanation, which they termed "radioactivity," in different uranium ores. Pitchblende proved to be an especially powerful emitter of radioactive rays and was a particularly promising starting point for their investigations. In 1898, after years of laborious work the Curies succeeded in isolating tiny amounts of two new, strongly radioactive elements from several tons of pitchblende: polonium and radium.

That same year, the pioneer Marie Curie also identified thorium as a radioactive element, although the German chemist Gerhard Carl Schmidt beat her by two months in publishing the discovery. (We will again encounter thorium and uranium with respect to stars and heavy element production.) For her research results, Marie Curie received two Nobel Prizes: in physics in 1903 together with Becquerel and Pierre Curie, as well as in 1911 in chemistry. This makes her the only person—while also being a woman—to have been awarded two separate Nobel Prizes in two different scientific disciplines.

Following the discovery of radium in 1899, Ernest Rutherford used it, and other elements, to examine its so-called α-radiation, which was still unexplained. In 1907, it was recognized that such radiation consisted of emitted "α-particles," which are equivalent to ionized helium atoms. We know today that "α-decay" is a spontaneous, radioactive disintegration of a nucleus emitting an α-particle. The original nucleus "decays" as a result, and the new nucleus has two fewer protons and two fewer neutrons. Interestingly, in 1907 nobody knew anything about "nuclei," "protons," and "neutrons" because they were identified as such only in 1910 (by Ernest Marsden, a collaborator of Rutherford), in 1919 (by Rutherford), and in 1932 (by Chadwick, a student of Rutherford), respectively. Apart from α-decay, Rutherford also observed more intense "β-decay," whereby an electron is emitted instead of an α-particle, as well as "γ-decay" through the emission of high-energy γ-rays. Rutherford likewise found that radioactivity led to the formation of new, lighter elements, and was the first to discover the law describing the half-life of radioactive material. In 1908, he received the Nobel Prize in chemistry—three years before Marie Curie. Twenty years later, in 1928, Gamov solved the problem of explaining the spontaneous α-decay of a nucleus by the quantum-mechanical effect of "tunneling."

Rutherford worked on more "practical" things, too. The Sun's energy source was an important and still unsolved problem. Around 1850, the German physicist Hermann von Helmholtz had recognized that energy is generally conserved. This important physical law states that energy can be neither created nor destroyed. Energy can only change its form, such as from potential energy into kinetic energy and then into thermal energy. An example would be an apple hanging on a tree, which first converts its potential energy, given by its height above the ground, into kinetic energy while falling down, and then into heat upon hitting the ground. Likewise, the energy radiated by the Sun has to be converted from another form of energy before being released.

Before 1900, the British physicist Lord Kelvin (William Thomson) as well as the German von Helmholtz (incorrectly) assumed that the pressure in the Sun's interior would decrease as it steadily cooled down, causing it to contract slowly under its own weight. The conversion of potential energy into radiation energy was supposed to be the source of the Sun's energy. It was soon shown, though, that this kind of energy production did not suffice for the Sun to keep shining for more than 20 million years. It contradicted biological and geological findings indicating that Earth itself was at least 300 million years old, if not even up to a billion years old. In 1904, Rutherford proposed that the Sun's radiation would be attributable to a special internal source of energy. He was probably thinking of radioactive decay inside the Sun. His research had confirmed that elements can emit radiation and thereby release around one million times more energy than any chemical reactions. But not even a fully radioactive Sun could produce enough radiation to solve the problem of the stellar energy source.

It was against this backdrop that around 1916 the English physicist Arthur Eddington began to work on the internal structure of stars and their evolution over time. He was interested in explaining variable stars and wanted to understand the energy sources of stars. He devised the first theory of the physical processes within a stellar interior that differed radically from what had been known before. The German physicist Karl Schwarzschild had already completed several studies on radiation pressure, which Eddington used and supplemented. These models described a star as a ball of gas that is prevented from gravitational collapse by thermal pressure. Eddington's important contribution was to show that radiation pressure is necessary to keep the stellar sphere in equilibrium.

Evidently, it was neither radioactivity nor gravitational collapse that provided the Sun with its energy. Eddington was familiar with Einstein's papers on the equivalence of mass and energy, and although many scientists were skeptical of this new concept, Eddington saw therein a possible solution to his problem. In his view, it provided the only explanation for the Sun's enormous amount of radiation. A fusion of hydrogen into helium seemed to be the sole possible energy source. In this process the most binding energy is released. The reason is that the sum of two protons and two neutrons is somewhat heavier than a single helium nucleus. This effect is known as mass defect. Applying Einstein's formula $E = mc^2$, it should therefore be possible to measure a "loss of mass" in the form of an energy discharge as helium is created in stars. Nevertheless, there were a few fundamental problems with this concept: (1) At that time it was still unknown that stars are mainly composed of hydrogen. How could there be enough fusion of hydrogen into helium if there was presumably hardly any hydrogen in stars? (2) The idea of fusion itself was questionable. Two hydrogen atoms are positively charged protons and consequently mutually repel each other. Eddington calculated that the temperature in the Sun's interior would need to be at least 40 million degrees in order for the protons to overcome their repulsion. The Sun's interior was thought to be far less hot than that!

Despite these setbacks, Eddington kept believing in his theory of "transmutation," the fusion of hydrogen and its serving as the energy source. He applied his new stellar model to calculate the temperature, density, and pressure at each location inside a star. He could even show that the temperature in a star's interior must be millions of degrees. In 1924 Eddington was still confident in the usefulness of his model because he could predict a mass-luminosity relation for stars. He was so convinced of the importance of this prediction that he publicized his model despite the fact that the underlying physics was still largely unknown.

Not everyone took kindly to the idea.

Although one of his fellow English colleagues, James Jeans, did not believe in Eddington's model, he suggested that stellar matter might be mostly ionized. This turned out to be an important improvement of the model. All the same, Jeans and other scientists continued to adhere to the older Kelvin-Helmholtz mechanism of energy production because it was based on classical mechanics. Eddington's revolutionary suggestions and new ideas, in contrast, dealt with the consequences of nuclear reaction processes that went far beyond classical mechanics. Despite an incomplete explanation, in the end Eddington's model evolved into an important tool in stellar astrophysics that made it possible to calculate the evolution of stars, for instance. He published his theory in 1926 as *The Internal Constitution of Stars*, which became a standard reference book for astrophysicists for many years.

Around the same time in Germany quantum mechanics was further refined by Max Planck. He carried out this work in the university town of Göttingen, where the emigrant Russian physicist George Gamov also spent some time. Gamov was especially interested in the concept of nuclear fusion but looked at the problem from an entirely different angle. He studied how radioactive elements lose their protons during an α-decay. If protons can somehow leave the nucleus, the process might as well work the other way around too.

In 1928, Gamov introduced the liquid drop model to describe nuclei, which was later improved by Niels Bohr and others. It describes the nucleus of an atom as a drop of incompressible "nuclear fluid." Consisting of protons and neutrons, it is held together by the strong nuclear force.

In accordance with the laws of classical physics, it should be impossible for nuclear particles to overcome the strong nuclear force that binds the nucleus together, to escape from it. Neither should it be possible for two protons to overcome their mutual electric repulsion and fuse into a new nucleus at stellar temperatures. But then, protons do fuse after all. The explanation for these processes is provided by the tunnel effect of quantum mechanics, described for the first time by the German physicist Friedrich Hund in 1927. It outlines the non-negligible probability that, owing to Heisenberg's uncertainty principle, a nuclear particle is able to overcome the barrier of a repulsive force despite having insufficient energy and to simply tunnel its way through. Gamov used this effect to explain radioactive α-decay, and how a nuclear particle can overcome the strong attraction of the nuclear binding force. Later, Gamov and Max Born independently applied the tunnel effect to explain the fusion of two protons.

In 1929, upon Gamov's advice, the German physicist Fritz Houtermans and the Briton Robert Atkins applied the tunnel effect to stellar energy production. They quickly showed that Eddington's ideas about nuclear fusion as an energy source were correct. According to their calculations, temperatures of "just" 40 million degrees would suffice to make occasional nuclear reactions possible by virtue of the tunnel effect. (Based on more advanced calculations it was later shown that temperatures of just a few million degrees are actually enough.)

They calculated reaction rates for the fusion of a whole range of nuclei without knowing which reactions are, in fact, feasible inside stars. They found, though, that only if hydrogen was participating in the reactions could enough energy be released. One thing became clear: nuclear fusion was occurring in stars and supplied enough energy for them to shine for long periods. Still, many details were unknown: What reactions are really involved? How much energy do they produce? And exactly how can two protons fuse to become helium?

It was years later before those still unanswered questions were asked again. Nazism cost Germany many of its best scientists, who left for other countries such as the United States, which quickly became the new center of scientific progress. The events in Europe slowed everything down, making it difficult for new ideas to flourish.

Carl Friedrich von Weizsäcker, a young physicist who had decided to stay in Germany, was greatly interested in the nuclear processes in stellar interiors and more generally, the binding energies of atomic nuclei. Given the discoveries of protons and neutrons—the components of a nucleus—and awareness of the tunnel effect, it was possible, at last, to compute the binding energies of the different kinds of nuclei. His research in this area soon produced fundamental results. In 1938, he published the first detailed calculations on the energy production through the fusion of hydrogen into helium as part of the so-called carbon (-nitrogen-oxygen) cycle (CNO cycle), whereby the heavier CNO elements act as catalysts. By using the CNO cycle, von Weizsäcker managed to circumvent the problem that without neutrons it is impossible for two protons to fuse directly into a two-proton nucleus. The enormous mutual repulsion between protons would cause a two-proton nucleus to immediately fall apart into two individual protons.

Entirely independent of von Weizsäcker's work, another nuclear physicist in the United States, Hans Bethe, was also interested in the energy source of stars. In 1939, in collaboration with Charles Critchfield, one of his students, Bethe began to work on possible fusion mechanisms. He suggested that β-decay could convert one of the protons into a neutron. Then, the result is not a two-proton nucleus but deuterium ("heavy hydrogen," a nucleus with one proton and one neutron), as the first product in a whole chain of reactions that eventually leads to the formation of helium. This solution made it possible for him to describe the proton-proton (p-p) chain of reactions through which hydrogen can be fused into helium.

As soon as the reaction rate for the production of deuterium was calculated using quantum mechanics, the energy output could be determined. When compared with measurements of radiation from the Sun, it became evident that the p-p chain reactions were a possible source of its energy. Bethe also examined the CNO cycle because energy production by the p-p chain is inadequate for stars hotter and more massive than the Sun. Since more massive stars have higher reaction rates, he could show that the CNO cycle is their main source of energy. Hence, both these processes were responsible for stars to shine throughout their long lives. The problem was finally solved: stars shine as a result of very specific nuclear reaction processes in their interior!

Nuclear astrophysics was born. Scientists now studied nuclear fusion and its role in the synthesis of the chemical elements inside stars. In 1967, Bethe was awarded the Nobel Prize in physics for his important contributions to stellar nucleosynthesis. Yet, around 1939 neither Bethe nor von Weizsäcker could explain how the formation of heavier elements occurs. Bethe proposed that elements heavier than helium, such as carbon, should have existed before stars formed. He was convinced that the production of neutrons in the interior of stars was negligibly small. As we know today, all heavier elements are successively built up in various neutron-capture processes and for that reason need enormously powerful stellar neutron sources. However, this insight only came about more than 15 years later.

2.5 Modern Alchemy

The knowledge that large amounts of energy are generated in stellar interiors through the nuclear fusion of hydrogen, the lightest of all elements, revolutionized the understanding of the cosmos. But the heaviest elements had also been subject to a lot of research since 1900, when Becquerel and the Curies had advanced the understanding of radioactivity. Around 1905, the Austrian physicist Lise Meitner and the German chemist Otto Hahn also began to take an interest in this topic. When nuclear fusion was described around 1938 and 1939, other new analyses of radioactivity led to the speculation that heavier elements could also release energy—albeit through nuclear fission, not through fusion.

In 1907, Lise Meitner began working with Hahn at the Kaiser-Wilhelm-Institute for Chemistry in Berlin, the start of a 30-year-long collaboration, with each of them heading their own department at the institute. This physicist/chemist duo each brought their unique expertise to their joint exploration of the field of radioactivity and the properties of the heaviest elements.

Hahn was more of an experimentalist, whereas Meitner worked mostly theoretically on the underlying physics. In 1918, they discovered the long-lived radioactive element protactinium (atomic number $Z = 91$). In the decay chain of the heavy radioactive element uranium-235, protactinium is produced even before actinium ($Z = 89$). Hahn then discovered the first known nuclear isomer of uranium in 1921, while he was analyzing the decay series to which this naturally occurring element belongs. Isomers are nuclides with equal numbers of protons and neutrons, but they are in different, long-lived states. Hahn and Meitner also experimented with neutron bombardments on uranium and thorium. They wanted to learn about new elements supposed to be heavier than uranium. Ever since the discovery of neutrons in 1932, it was presumed possible to make such transuranium elements in the laboratory. A rivalry ensued over the possibility of winning a Nobel Prize for such work, and a race began between contenders Hahn and Meitner in Germany, Rutherford in England, Irene Joliot-Curie in France, and Enrico Fermi in Italy.

Around 1930, the chemist Fritz Strassmann joined the Hahn-Meitner team in Berlin. He assumed Meitner's tasks in 1938 when she, being Jewish, lost her university post and had to flee to Sweden. Hahn and Strassmann performed the experiments that they had initiated with Meitner before her involuntary departure. Hahn continued to regularly write to Meitner to inform her of their results. In Berlin her two colleagues indeed succeeded in observing the first experimental signs of nuclear fission. Their bombardment of a uranium nucleus with neutrons unexpectedly produced the element barium ($Z = 56$), which is only about half as heavy as uranium, even though they had actually been hoping to produce heavier nuclei.

Hahn described this new reaction to Meitner as a "bursting apart" of the uranium nucleus and asked her for some "fantastic explanation" of this inexplicable process. The new experimental findings were published in January 1939 without a mention of Meitner because Jews were not allowed to publish in Germany anymore.

Meitner's nephew, the physicist Otto Frisch, was visiting her in Sweden at the end of 1938. He was there for Christmas when Meitner learned of the results of Hahn's recent experiments. During a walk in the snow, Meitner and Frisch continued to discuss how the uranium nucleus could have split like a droplet of water, into two nuclei of similar size. The detected barium would therefore have been one of those two new nuclei.

Meitner and Frisch then calculated that the process was, in fact, energetically possible. The mass defect describes the difference of the total mass of these two new nuclei compared to the mass of the original uranium nucleus. Using Einstein's formula that includes the mass defect, they found the same energy is produced by the repulsion of the two new equally charged nuclei after the uranium nucleus' unintentional splitting.

The conclusion that was drawn was that the nuclear fission of just a few kilograms of uranium ought to produce the same explosive power as many thousands of tons of the chemical explosive trinitrotoluene (TNT). In February 1939, Meitner and Frisch published the now famous theoretical physics explanation of those experiments. They called the newly discovered process "nuclear fission." In the meantime, Frisch had already succeeded in directly isolating the fission products and was able to provide the experimental proof himself. His result was published the following week. Not long afterward, this finding was verified worldwide.

Scientists had been examining what they thought to be nuclear fusion and discovered nuclear fission instead. They wanted to discover new elements and had found a new energy source. They were intensely engaged in basic research and unsuspectingly changed not only our understanding of the Universe but also, very dramatically so, life on Earth.

In 1944, Hahn received the Nobel Prize in chemistry for his research on nuclear fission—without Meitner and Strassmann. Her explanation of nuclear fission quite unfairly remained unacknowledged. She was, however, awarded the important Enrico Fermi Prize together with Hahn and Strassmann in 1966.

Today there are many prizes and lecture series named after Lise Meitner in honor of her achievements. I, too, was invited to deliver two Lise Meitner Lectures in 2010 under the auspices of the German and Austrian Physical Societies. I was humbled to follow in her footsteps and report on the origin of the elements as well as the dating of ancient stars by means of the natural radioactive decay of the long-lived isotope uranium-238. Speaking about footsteps, it is a nice coincidence that during the 1960s Hahn was living in the same neighborhood in Göttingen as where my mother grew up. She can still remember very well occasionally passing the famous man on his afternoon walks. I have visited my grandparents there many times, but I never imagined that I would one day be writing about Hahn and his discoveries.

2.6 The Foundation of Cosmology

The existence of a Universe full of other galaxies beyond the Milky Way had become common knowledge since 1925. This led to the development of many theories about the cosmos as a whole, in other words, about cosmology. There had already been many philosophical considerations about this topic, but only then was it possible to formulate more profound physical theories that could also be supported by observations.

In 1916, Einstein's research culminated in his general theory of relativity, which was a consistent description of gravitation as a geometrical characteristic of time and space. Applications of his theory of relativity offered a new opportunity to view the Universe as a whole and to describe it mathematically. It was the beginning of modern cosmology. To solve his mathematical equations Einstein himself initially imagined a static Universe, which agreed well with the assumptions of his day.

According to this solution, the Universe should neither grow nor shrink over time. To avoid collapse under its own gravity, Einstein needed another force to counteract gravity in his equations, and so the long debated cosmological constant was born.

Shortly afterward, in 1917, Willem de Sitter published his ideas about the formation and evolution of the Universe. His model was a bit obtuse and hard to comprehend because it did not contain any matter. Nevertheless, a test particle could recede from an observer leading to the prediction of a redshift for distant cosmic objects. Astronomers immediately sought observable proof of de Sitter's model by looking for nebulae and galaxies showing a large redshift.

During the next 10 years cosmology evolved slowly, but stellar observations and spectral classifications were already under way, and quantum mechanics was being established. The first results were obtained with respect to the search for the energy source of stars by the mid-1920s, when the next piece of the cosmological puzzle came about.

It did in an unlikely and obscure figure, the Belgian priest Georges Lemaître, who began to develop a serious interest in astronomy and the Universe. He had studied physics under Eddington in England, with whom he would later, over the course of many years, be discussing cosmological models and the evolution of the Universe. Their collaboration was a consequence of Eddington's regular meetings with de Sitter at the conventions of the Royal Astronomical Society in Britain, where the two discussed Hubble's observations from 1923 of distant redshifted galaxies trying to find a cosmological interpretation for them.

Around 1930, Lemaître read a report about those meetings and wrote to Eddington to inform him of ideas and works that he had developed before 1927. These were building on the work of the Russian Alexander Friedmann, who was also deeply interested in cosmology. In 1922, he had developed a solution that described the Universe first as expanding, and then as contracting without the need for any cosmological constant. Such a cyclical Universe that expands and then shrinks again conveniently avoided the problem of the unknown origin of the Universe. Later, in 1924, Friedman also found another solution for a forever expanding Universe, but Einstein's field equations still could not be solved for the general case. In 1933 the English astrophysicist Edward Arthur Milne had simplified Einstein's equations into a so-called

Friedman equation by introducing the cosmological principle, which considers the Universe homogeneous and isotropic.[1] Lemaître had found a solution to this simplified equation in which the Universe is uniformly expanding. Thus Lemaître came to be one of the first people to grapple with Einstein's general theory of relativity and to apply it to the Universe as a whole.

From the expansion of the Universe, Lemaître also derived a linear distance-velocity relation for cosmic objects. He considered this law a consequence of relativistic cosmology. Furthermore, Lemaître calculated a value for the "Hubble constant," which describes the expansion rate of the Universe. However, he was not able to recognize this relation in the astronomical distances of cosmic objects and their velocity measurements, as there simply were not enough observational data available in 1927.

All in all, unnoticed by the rest of the world, Lemaître was the first to conclude from contemporary theories and galaxy observations that the Universe expands. If the Universe is expanding, the next question would be about its origin. Lemaître and Eddington continued to consult each other regularly about this problem and its consequences in the years that followed. Lemaître's idea of an expanding Universe had laid the cornerstone for the current theory describing the Big Bang as the origin of our Universe. Only in retrospect did it become apparent how very much Lemaître's new Big Bang theory would change the contemporary worldview of a static Universe.

Hubble discovered the distance-velocity law of galaxies two years later, in 1929, on the basis of years of systematic observations of distant galaxies taken by him personally as well as by Milton Humason and Vesto Slipher. This new law describing the expansion of the Universe made him famous around the world. The name Hubble is familiar to many due to the space telescope that was named after him in 1990. Lemaître's papers, on the contrary, were originally published in French in

---

[1] "Homogeneous" denotes that the same observational results are to be expected from different locations in the Universe. Consequently, any part of the observable Universe yields a representative result. "Isotropy" means that from all viewpoints the same observational results are to be expected. Thus, the same physical laws apply everywhere.

a rarely read Belgian journal accessible to just a few of his contemporaries. The American mathematician and physicist Howard Percy Robertson had also already used an expansion rate in his cosmological computations of 1928, independently of both Lemaître and Hubble. However, the discovery of the expansion rate is most often attributed to Hubble alone.

Motivated by his conversations with Eddington, Lemaître further developed his ideas about the origin of the Universe. Radioactivity probably inspired him to introduce in 1931 the idea of a "primeval atom," from which the entire Universe supposedly formed as it gradually underwent "decay." He also postulated that the start of this decay was the beginning of space and time. There was a significant problem with Lemaître's model, though. The age of his Universe, then thought to be two billion years, did not agree with the considerably older age of the Sun. In the meantime, Eddington had also developed his own model of an initially static and subsequently expanding Universe, and although he could not describe the transition between these two phases, age did not pose a problem for his model.

In 1931, even when Einstein rejected the cosmological constant after the discovery of the expansion of the Universe, neither Lemaître nor Eddington could be dissuaded from continuing to use it. On the contrary, Lemaître refined his theories of an expanding Universe in 1933, ultimately making him a pioneer of modern cosmology. He interpreted the cosmological constant as being a result of a "vacuum energy" with a perfect equation of state. It was a wise intuition on his part, as was only much later discovered.

In the 1990s two groups of researchers headed by Saul Perlmutter and by Brian Schmidt and Adam Riess discovered that the Universe is not just expanding but actually accelerating. In 2011, this discovery was rewarded with the Nobel Prize in physics. The accelerated expansion can be described mathematically with the cosmological constant. The underlying physical meaning, unfortunately, is still unclear. This perplexity is also reflected in the name of "dark energy," which is supposed to cause the acceleration. At least we know that dark energy contributes 72% to the total energy budget of the Universe. Some three generations later Lemaître was proven right.

While all these debates about the origin and evolution of the Universe were under way in Europe, the Swiss astronomer Fritz Zwicky in Southern California was systematically observing the galaxy cluster Coma Berenices (Bernice's Hair) in 1933. He was interested in measuring the redshifts of the individual galaxies within the cluster. However, they were moving so rapidly that they should have been escaping from the cluster, assuming that the gravitational effect corresponded to his observations of the cluster's luminosity or mass. Something was wrong because the Coma cluster did not look as if it was losing all of its galaxies. Zwicky hypothesized that quite a lot of non-luminous matter in the galaxy cluster would be holding it together by its additional gravity. Zwicky's ideas were soon forgotten, however, because not enough proof had been supplied at the time. Today we know that "dark matter" is involved, just as Zwicky had surmised.

Systematic observations of galaxies were only carried out toward the end of the 1970s by the American astronomer Vera Rubin. Around 1964, she was the first woman to be granted official permission to operate telescopes on Mount Wilson near Los Angeles. Theoretical work on the existence of dark matter around galaxies had already been carried out by the American astrophysicists James Peebles and Jeremiah Ostriker. Meanwhile, Rubin kept on determining the rotation curves of individual galaxies. She found a clear discrepancy between calculated values and her observations. She was thus able to show that galaxies possess at least 10 times more dark matter than luminous stellar and gaseous matter. To express this difficult-to-understand discovery differently, each galaxy is composed of at least 90% dark matter. The immediate conclusion followed that the visible Universe is also just a small percentage of the Universe as a whole. These findings were rapidly accepted, and the insight that dark matter is an important component of galaxies was established. The overwhelming majority of scientists today are convinced that this dark matter postulated by Zwicky does, in fact, exist. That said, the physical properties of dark matter remain unclear. It was a fundamental discovery for the study of galaxies and cosmology that would motivate enormous progress.

After the 1930s came a decade-long dry spell of new cosmological models, as the advance of science slowed down and took unfortunate turns. During that period it was uncovered how fusion makes stars shine, but it was also discovered how nuclear energy could be released on Earth through the fission of the heaviest nuclei. World War II and the Manhattan Project left their mark on both human history and scientific progress overall. New advances in cosmology were made again in 1948. The British mathematician Hermann Bondi, astrophysicist Thomas Gold and astronomer Fred Hoyle suggested a new model based on the so-called perfect cosmological principle. It postulates that the Universe is not just spatially homogeneous and isotropic but also temporally invariant. Thus, it not only looks the same in all directions over great distances but remains unchanged over time as well. The concept of a "steady state" model came to be commonly used to describe spatially and temporally invariant cosmological models. The model by Bondi, Gold, and Hoyle did not require Einstein's cosmological constant. But if the Universe is expanding, new matter needs to be forming constantly, otherwise it would be unable to satisfy the perfect cosmological principle. Eddington's and Lemaître's models, by contrast, were not homogeneous because matter originated just during the Big Bang and was gradually diluted due to the expansion of the Universe.

During a radio interview in 1950, Fred Hoyle explained how the rivaling theory began with a "big bang," a term that stuck. The two contending theories then were the Big Bang theory and the steady state theory. How could a decision be reached about which one was right? The discovery of the cosmic background radiation in 1965 by Penzias and Wilson just one year before Lemaître's death was not consistent with the steady state model. That model could not account for a hot initial phase leaving behind radiation still measurable at the present time. Therefore, the Big Bang hypothesis became widely accepted.

The age problem of the Universe was also finally solved in 1980. The American particle physicist Alan Guth and the Russian American cosmologist Andrei Linde independently made modifications to the Big Bang theory to introduce an early inflationary period during which the Universe expanded extremely rapidly for an extremely short period. Since then, various predictions of this inflation theory have been observationally confirmed such as with data from the WMAP satellite. Inflation is, as a consequence, now a part of the hot Big Bang model.

There is a third observation in support of the Big Bang theory. According to it, the chemical elements hydrogen and helium form during the first few minutes at a ratio of three-fourths to one-fourth, including traces of lithium. This is congruent with observations of gas that has remained unchanged since the Big Bang.

2.7 The Origin of the Elements

Let us now return to the processes that operate inside stars. Since the end of the 1930s, it was clear that inside stars hydrogen is converted into helium and that the stars acquire their energy from this fusion process. The origin of all the other elements remained to be clarified, though. Initially, it was assumed that they were not the products of stars. But where else could they possibly come from? The search for an answer began in 1946 with Fred Hoyle. He was the first to explain that, in principle, all elements in the periodic table could be synthesized inside stars. He suggested that newly generated elements were deposited back into interstellar space by supernova explosions, that is, during these spectacular explosions of extremely massive stars that ended their lives. These ideas were quickly adopted by others and developed further.

In 1948, three physicists, Alpher, Bethe, and Gamov, published a paper—somewhat jokingly referred to as the "αβγ-article"—as well as an additional detailed report written by just Ralph Alpher, proposing that the heaviest chemical elements could be formed by rapid neutron-capture.[2] They were still assuming that this kind of element synthesis took place shortly after the Big Bang with neutrons coming from the primordial material. Bethe had been convinced that no significant neutron source existed inside stars. We know today that the Big Bang nucleosynthesis depicted by Alpher, Bethe, and Gamov was inaccurate. However, their idea with regard to the synthesis of the lightest elements

---

[2]Nuclei can capture neutrons when they are bombarded with neutrons. Some of the captured neutrons convert into protons through β-decay. This way, a nucleus of another element is formed. Depending on the neutron density, the process is called r-process (for "rapid") or s-process (for "slow"). The r-process proceeds at high neutron densities, whereas the s-process results from low neutron densities. It is described in greater detail in chapter 5.

during the Big Bang, namely hydrogen, helium, and lithium, agreed with observations.

Since life is based on carbon compounds, Hoyle continued to think about the cosmic origin of this and other elements. Carbon had come from somewhere. He was not alone in this quest. The Baltic astronomer Ernst Öpik and the Australian American astrophysicist Edwin Salpeter independently postulated in 1951 and 1952, respectively, a threefold fusion of the helium nuclei, which is termed the "3α-process." In a first step, two helium nuclei were supposed to combine into a very rapidly decaying beryllium nucleus. Salpeter could show that at very high temperatures the pertinent fusion reaction for the beryllium production occurs somewhat faster than the subsequent decay. The outcome is equilibrium at a ratio of one beryllium nucleus to one billion helium nuclei. In a second step, the beryllium was then supposed to combine with a third helium nucleus to form carbon.

Hoyle criticized these calculations in 1953 based on arguments that the observed abundance of carbon could never have been produced at the temperatures of about 100 million degrees in stellar interiors as proposed by Salpeter. An exception would be if a resonance occurred in the carbon nucleus precisely at the energy of the second fusion reaction. Only such a resonance could drastically accelerate the second reaction. Later, in 1957, a resonance of this kind was indeed found by the American nuclear physicist William Fowler and colleagues, who were carrying out nuclear physics experiments.

This is an impressive example of a prediction of the characteristics of an atomic nucleus made purely based on astronomical observations and assumptions. Incidentally, the rapid decay of the beryllium nucleus is the reason why the 3α-process could not operate during the Big Bang. By the time helium nuclei were forming, the Universe had already cooled down too much, meaning that the fusion into beryllium would be considerably slower than its decay, preventing beryllium from accumulating.

In 1954, even before the resonance level had been verified, Hoyle introduced another fundamental concept regarding the synthesis of heavier elements. He described possible fusion reactions in the interiors of evolved stars to produce the elements from carbon to nickel. Nonetheless, in 1957, Gamov insisted once again that all elements had formed in fixed proportions during the Big Bang. In stars, only hydrogen could fuse into helium. Therefore, he concluded, no chemical evolution could occur following the Big Bang. This view, however, contradicted some observations of stars, although in those days this fact was not yet fully recognized.

The discussion started to change as another publication appeared, jointly put together by four scientists who again placed the origin of the chemical elements within stars. Margaret Burbidge, Geoffrey Burbidge, William Fowler, and Fred Hoyle, all of whom were in California (although Hoyle's university position was in England), had been working on the formation of the elements from beryllium to uranium. They agreed that hydrogen, helium, and lithium formed during the Big Bang and not in stars. Their results ultimately led to the theory of Big Bang Standard Nucleosynthesis, which remains accepted today. The famous paper on the "Synthesis of the Elements in Stars" combined, in over 100 pages, earlier findings with recent results. Not long afterward, it came to be regarded as the standard reference material for nucleosynthesis processes and element formation. Since then, when referring to the article, just the authors' initials "B^2FH" are used, and their theory is known as such around the world.

Observational evidence supporting the "B^2FH" theory came from the American astronomer Paul Merrill. He observed the radioactive element technetium (^{99}Tc) in asymptotic giant stars, which has a half-life of only 2.1 million years and thus relatively short-lived compared to stars with ages often in the hundred millions or even billions of years. Thus it immediately became clear that this element should have been produced in those stars themselves not long ago. (For completeness it should be noted that the longest lived isotope of technetium [^{98}Tc, with a half-life of 4.2 million years] is not actually produced in these stars.)

Around this time, in 1957, the Canadian nuclear physicist Alastair (Al) Cameron published his own calculations of stellar element synthesis. Cameron searched for the neutron source necessary for the production of technetium within the star. By calculating various reaction rates for the so-called slow neutron-capture operating in giant stars, he identified the reaction chains still considered valid today for the synthesis of elements heavier than nickel. Cameron devoted his entire life to explaining the processes of nucleosynthesis of the different lighter and heavier elements. Right up to his death in 2005 he was trying to track down the astrophysical locations of heavy-element genesis by means of sophisticated calculations. Many of those problems remain unsolved to this day.

By predicting that all heavier elements are synthesized in stars by various processes, B²FH and Cameron established the field of nuclear astrophysics that deals with the subject of stellar nucleosynthesis. One major advantage of this theory was the prediction of continuous chemical enrichment of the Universe, that is, a chemical evolution that no prior model had considered possible. Spectroscopy and the advent of large, modern telescopes made it possible to examine different kinds of stars to experimentally test the model, which led to its confirmation.

As explained by the well-established quantum mechanical theory astronomers know that different atoms emit or absorb specific wavelengths of light. Stellar spectra therefore inform us about the chemical composition of a star. Interestingly, some stellar observations indicated an anticorrelation between the age of a star and its abundance of heavy elements. The older the star, the smaller the amount of heavier elements it contains. This was baffling, at first, but could be explained by means of the newly developed Big Bang theory and Big Bang nucleosynthesis. Since only the lightest elements were produced in the Big Bang, it could be assumed that the very first stars in the Universe were composed exclusively of hydrogen, helium, and lithium. All the other elements then had to be gradually synthesized inside stars. Those new elements would later be deposited back into the interstellar medium by dying stars.

The next generation of stars was formed from gas slightly enriched with these heavier elements. Toward the end of their lives the stars expelled newly synthesized elements into interstellar space mainly by supernova explosions. The same then occurred for subsequent stellar generations. The resulting cycle explains why ancient stars that formed early on contain much smaller amounts of heavy elements than stars that were born later.

Both Cameron and B²FH went into great detail about the nuclear physics of nucleosynthesis as well as the astrophysics of the conditions under which element synthesis can occur. They managed to describe many different stellar environments in which characteristic processes can synthesize particular elements and isotopes. Specifically, they predicted a number of processes (e.g., the r- and s-processes of rapid and slow neutron-capture) that are responsible for the synthesis of elements heavier than iron and nickel.

Beginning in the 1970s, the use of increasingly powerful computers enabled more efficient and continuously improved nuclear astrophysics calculations. Until then, this field was relatively short of quantitative results. Among others, the American astronomer Donald Clayton developed preliminary time-dependent models of the slow and rapid neutron-capture processes. Earlier calculations had been performed only for constant temperature and density and could not properly account for different astrophysical environments of nucleosynthesis.

The process of nuclear fusion inside stars and the different stages of element nucleosynthesis during stellar evolution had been modeled for many decades since 1957. Models of massive stars that explode at the end of their lives predict the release of an immense amount of neutrinos during the explosion. The explosion is caused by the collapse of an iron core, the end point of stellar fusion reactions. The first such neutrinos were experimentally detected during the famous supernova 1987A observed in the Large Magellanic Cloud.

Calculations of the interior structure of the Sun were experimentally confirmed by two independent methods. The first method relied on the detection of neutrinos from the Sun's core, which revealed which nuclear processes, and at what rates, are operating in the Sun's core. The second method used the detection of sound waves to derive the pressure and temperature values for any position within the Sun, by analyzing how these waves pass through the gaseous solar sphere. This latter method is called helioseismology, an analogy to seismology on Earth, where earthquake waves are used to explore the Earth's interior.

My own research concerns ancient stars that contain only tiny amounts of heavy elements. They give us a unique opportunity to explore the early Universe, and to study the details of the diverse processes of nucleosynthesis. At the same time, each newly discovered old star is another small proof of the B²FH and Cameron theories. They remind us that we have not known for very long how and where the elements formed in the Universe: inside stars. We are indeed "children of the Universe," as the German physician and science journalist Hoimar von

Ditfurth first said in 1970. "We are made of star stuff," as the American astronomer Carl Sagan put it in 1973.

In 2007, a scientific conference was held in Pasadena, California, in honor of the 50th anniversary of the B²FH article. I attended this conference as a postdoc and presented a talk on spectroscopic abundance measurement of lead in a particular ancient star. I reported how those data, together with the calculation of lead based on the decay of the radioactive thorium and uranium over cosmic timescales could reveal new details about the direct production of lead in the process of rapid neutron-capture. This star continues to be an important part of my research and is also an excellent "laboratory" for nuclear astrophysics.

Margaret and Geoffrey Burbidge as well as Ed Salpeter were at the conference, although only Geoffrey gave a talk about the B²FH article and its continued influence to the present day. The Burbidge couple did not follow the entire conference program to avoid exhaustion, leaving only a few chances to make their acquaintance or even converse with them. However, I was able to meet Margaret Burbidge at the conference dinner on the fourth day of the conference. There I was, standing face to face with the 88-year-old grande dame of nuclear astrophysics: a modest women with a kind smile. I stuttered something about "a great honor to meet you" and "I'm also working on the rapid neutron-capture process." I believe she could not understand the half of it, owing to significant background noise. With the bustle of the conference dinner reception and the presence of quite a few other admirers, it was not possible to talk to her for more than two minutes. Overall I was deeply impressed, and my reverential muttering was not really important. To me it was just wonderful to have shaken her hand and to have seen who this Margaret Burbidge actually is.

Another special moment followed soon afterward when she agreed to pose for a photo of the two of us. It can be seen in Figure 2.5 and depicts a proud moment for me as a scientist. My research area directly follows the groundbreaking results by B²FH. They described the process of rapid neutron-capture, which I continue to investigate with my stars. Professionally speaking, I am the Burbidges' scientific granddaughter who keeps pursuing nuclear astrophysics "experimentally" with the observations of the oldest stars.

Figure 2.5. Margaret Burbidge and the author in 2007 at the "50 Years of Nuclear Astrophysics" conference in Pasadena, California, USA. (*Source*: Photograph by Anna Frebel. With the kind permission of Margaret Burbidge)

CHAPTER 3

STARS, STARS, MORE STARS

Stars need enormous amounts of energy to shine for millions, even billions of years. This energy is released by the processes of nuclear fusion inside them. How exactly do these processes operate in stellar interiors? Do they occur in the same way in all stars? And how much do those stars all resemble each other? These and many other questions concerning stellar physical processes show that stars are actually quite similar to people. Seen from far away, they all look more or less the same besides their brightness and color. But if you look a little closer, you will see that each has its own personality and consequently exhibits entirely individual features.

3.1 The Cycle of Matter in the Universe

A cycle of matter of truly cosmic proportions is the basis of the formation and evolution of all stars and galaxies in the Universe. The cartoon depicted in Figure 3.1 is a helpful first guide to this gigantic cycle.

The matter cycle began a few hundred million years after the Big Bang with the formation of the first stars. These massive objects emerged from gigantic clouds of gas that collapsed under their own weight to condense into stars. At that time, the Universe was still composed of the original primordial material, that is, around 75% hydrogen, 25% helium, and traces of lithium. After a short life of just a few million years, these first element manufacturers exploded as powerful supernovae. The new elements formed during those stars' lifetimes and during the explosions were expelled far out into space by the sheer power of those supernovae. The initially "unpolluted" primordial material was suddenly changed forever. Chemical evolution had begun.

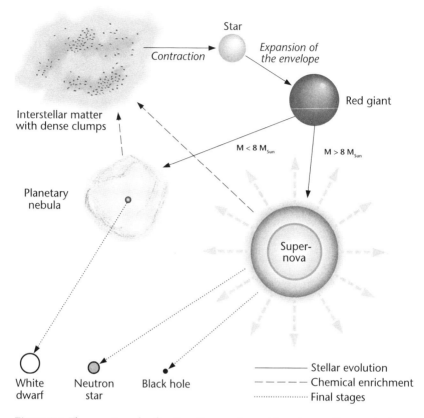

Figure 3.1. The cosmic cycle of matter: Stars are formed from interstellar gas. As they evolve, new elements are synthesized in their interiors, which are returned to the interstellar medium by either stellar winds or supernova explosions. For that reason each successive generation of stars possesses a somewhat higher content of heavy elements because they are formed from interstellar medium that had been increasingly enriched. (*Source*: Peter Palm)

The existence of new elements in the Universe had far-reaching consequences. The chemical and physical properties of gas, for example, depend on its chemical composition. The Universe at that point in time can be imagined as a huge cauldron. A clear broth is slowly simmering inside. If you took a taste of it, it would probably taste rather bland. What's missing? Some salt, of course. Elements heavier than hydrogen and helium are, for the Universe, what salt is for a soup. Although present in only the tiniest of quantities, they do change the taste and therefore the behavior of the gas.

The presence of those first heavier elements then set in motion the cycle of matter. The next generation of stars formed from a mixture of gas with a slightly different chemical composition than the one from which the very first stars formed. These new stars, therefore, were made not from primordial matter but "salted" gas.

Thanks to the new elements, the gas was able to cool enough for smaller and smaller parts of the gas cloud to collapse into separate stars. This way, stars much less massive than the colossal first generation of stars could form. To generate energy, all stars fused hydrogen and other elements into heavier ones in their interiors. The most massive stars of this second and all subsequent generations ended their short lives in enormous supernova explosions. Those explosions mixed the newly synthesized material back into the interstellar gas. This chemical enrichment process repeated itself from one generation to the next over the course of billions of years.

Supernova explosions of massive stars were not the only mechanism by which heavier elements were dispersed. There were also stellar winds that result from gas being continually blown off a star's surface into space. Throughout stellar evolution, that is, during a star's "lifetime," significant amounts of elements such as carbon can leave the surface of a star and make their way into the interstellar medium. The Sun also loses small amounts of surface material through its solar wind. We can witness this in the form of the Northern Lights. This mass loss is initially insignificant for most stars, except for the most massive ones. The loss of surface material can even alter the fate of a star in later stages of its evolution. If enough mass is lost, a previously massive star will, for example, not end as a supernova anymore.

More generally, low-mass stars will expel almost their entire hydrogen envelope. In some cases, a star's expanding envelope is directly observable streaming away: this is called a planetary nebulae. All the new elements from the core region that had been mixed into the envelope during earlier phases of stellar evolution are returned to the interstellar medium with the loss of the envelope. The remaining core of the star is called a white dwarf. It consists mostly of helium, or carbon and oxygen.

No nuclear reactions occur in white dwarfs, as opposed to in stars. They simply cool down until they become as cold and dark as the Universe itself. If a white dwarf is part of a close binary-star system, the story can end up entirely differently, though. If the white dwarf becomes heavier than 1.4 times the mass of our Sun, as a consequence of receiving mass from a binary companion, it will collapse under its own weight, which leads to the explosion as an energetic supernova. This explosion is triggered by a sudden onset of uncontrolled nuclear reactions that completely transmute the original material of the white dwarf into heavier elements up to iron. Those newly generated elements are then returned into the interstellar medium during the explosion.

Not every star or supernova generates the full set of all elements from the periodic table. This "task" is divided between the different types of stars, their stellar evolutionary phases, and explosion mechanisms. The various processes of nucleosynthesis in combination with stellar properties such as their mass then play a crucial role in determining which elements will be synthesized and in what quantities. In most cases, groups of neighboring elements with sequential atomic numbers in the periodic table are synthesized, as is described in detail in section 3.3.

For technical and other reasons based on atomic physics details, not all the elements present in a star's outer envelope are measurable. However, it is possible to detect up to 65 elements in various types of stars depending on their surface temperature and other properties as well as the quality of the data. A special case is the Sun. A total of 83 elements have been detected, 64 through so-called spectral analysis and 19 others from meteorites assumed to have formed from the same gas cloud as the Sun.

Today stars continue to steadily and reliably drive the matter cycle forward. They are responsible for the cosmic production of elements heavier than hydrogen and helium. As time passes, the total content of these elements steadily rises in each galaxy as well as in the Universe as a whole. Consequently, each new generation of stars contains a tiny bit more of the heavier elements than the preceding one. Their increasing amount is due to the fact that newly synthesized material cannot be extracted from the cycle. The only way to remove larger amounts of elements would be through the formation of compact stellar objects, such as white dwarfs, neutron stars, black holes, or even planets, because they simply lock up material. But in the end, the balance is tipped far in favor of continued element production. More details on all these processes will be given in the following chapters. The Sun reflects this chemical evolution: It contains the relatively large amount of 1.4% of all the elements because it was born "just" 4.6 billion years ago from gas already enriched by many—likely thousands—of generations of stars.

We human beings have a long evolutionary history, not only biologically on planet Earth but also cosmo-chemically, with the evolution of the elements in our Galaxy. This chemical evolution prepared for the existence of the Sun, Earth, and ultimately also life itself. Carl Sagan's message that "we are made of star stuff" fittingly reminds us that we are part of chemical evolution, along with the Sun and our Solar System. As the descendants of the stars, we carry their cosmic genes inside ourselves, namely, the chemical elements.

3.2 Astronomers and Their Metals

In astronomy, it is often of great importance to describe as many facts and characteristics of the cosmos in the simplest possible fashion. After all, the Universe is complicated enough as it is. One example is the chemical composition of a star. To describe it, a short and simple notation was devised a long time ago: X denotes the amount of hydrogen in a star and Y the amount of helium. All the remaining elements are summarized as Z. Furthermore, in astronomy, all the elements heavier than hydrogen and helium are called "metals." X, Y, and Z can then be used to compile the "astronomer's periodic table," so to speak. It is shown in Figure 3.2.

Any chemist is likely going to roll her or his eyes hearing the term "metals" because in chemistry not all elements are metals, by a long shot. But every field has its idiosyncrasies. Astronomy surely is one of the quirkier disciplines in this regard since it still is heavily influenced by historical classifications, notations, and customs. New definitions often gain acceptance only very gradually.

The chemical composition of a star can be characterized by information about X, Y and Z. One typical example for X, Y, and Z is

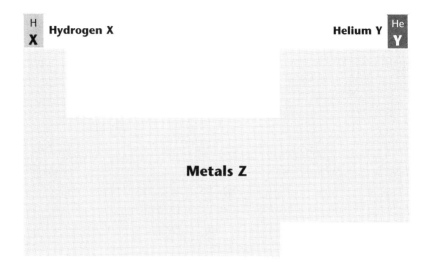

Figure 3.2. The astronomer's periodic table. Three things are important in the Universe: hydrogen, helium, and metals. (*Source*: Peter Palm)

provided by our Sun: X=0.715, Y=0.27, and Z=0.014. As a star is formed from an interstellar cloud of gas, its composition thus corresponds to the composition of the cloud, which itself mainly constitutes hydrogen and helium. This explains why the amount of metals, hence Z, is always extremely low compared to the star's total mass. Values for Z range from Z=0.000001 to Z=0.04, depending on when the star formed.

Despite this apparently negligible fraction of metals, Z signifies a fundamental measurable property of a star. It provides crucial information for classifying it chemically because, in the end, it sets all stars apart. Just as Kirchhoff, Bunsen, and even Huggins over a century ago, astronomers today also want to determine the chemical composition of the stellar atmospheres and which elements are present in the stars. What an observational astronomer can determine is not the mass fractions for X, Y, and Z, but the number densities of the atoms of the corresponding elements. In principle, the masses can also be calculated from the atomic weights of the elements and the density of the gas. In practice, however, the gas density is not known precisely enough.

Fortunately, the values for X and Y, as measured at the stars' surface, hardly ever change in most stars. The exception is shortly before the end of their lives. It is only after stars eject their outer hydrogen envelopes at the end of their evolution that the lower-lying hotter gas layers appear in which hydrogen had fused into helium. The inner layers of a star then become observable, whereby the value for X reduces dramatically, and accordingly, that of Y increases. Since all stars possess about the same amounts of X and Y throughout their evolution, such drastic changes very clearly indicate the end of a star's life.

The amount of metals, Z, can be determined for many stars, and chapter 7 describes those methods in detail. This quantity is also called metallicity. The metallicity of a star depends, above all, on when it was born. As can be seen in Figure 3.3, the Universe was still "metal-poor" during its early epochs. At that time, any elements present had been supplied by just a few generations of supernovae. Accordingly, stars with extremely low abundances of metals formed from such metal-poor gas. They are called metal-poor stars.

The concept of a "metal-poor star" is a basic, very commonly used term in stellar astronomy. Stars with a metallicity lower than one-tenth of the Sun's metallicity are generally called metal-poor. In many cases, we shall be talking about stars that are substantially more metal-poor, so-called extremely metal-poor stars. They have less than one-thousandth of Z compared to the Sun. This means that only 0.001% of

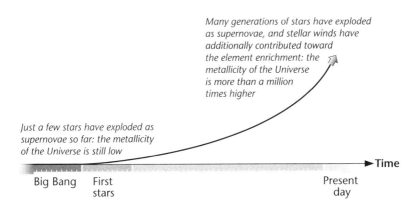

Figure 3.3. Schematic view of chemical evolution. Since shortly after the Big Bang, primordial matter has been successively enriched with newly synthesized metals by generations of supernovae and stellar winds. The most metal-poor stars formed in the early, still metal-poor Universe. (*Source*: Peter Palm)

Table 3.1. Definitions of Metal-Poor Stars

Type	Definition
Sun	Reference star
Metal-poor	1/10th of the solar iron content
Very metal-poor	1/100th of the solar iron content
Extremely metal-poor	1/1,000th of the solar iron content
Ultra metal-poor	1/10,000th of the solar iron content
Hyper metal-poor	1/100,000th of the solar iron content
Mega metal-poor	1/1,000,000th of the solar iron content

Source: Data from Beers and Christlieb, *Annual Review of Astronomy and Astrophysics* 43 (2005): 531–580.

the atoms in those stars are heavier than hydrogen and helium. Table 3.1 lists the definitions of the various metal-poor stars to illustrate current metallicity classes. As will be described in later chapters, to account for the recent discovery of an iron-poor star, a new definition for objects containing one-ten-millionth or less of the solar iron content needs to be added.

In principle, it is possible to conclude whether a star is younger or older simply based on its metallicity: the lower the metallicity, the earlier the star formed. The Sun serves as the reference star and defines the "zero point," even though it is already about 4.6 billion years old. Therefore, younger stars do exist with a higher metallicity than that of the Sun. They are often described as super metal-rich, and the highest known value is about three times the solar value. The overall lowest known metallicity is now less than one-ten-millionth of the Sun's value. These most metal-poor stars are the messengers from the early Universe that lead us back to the period shortly after the Big Bang.

The success of stellar archaeology rests on the basic assumption that metal-poor stars formed from gas in the early Universe and that this gas has remained unchanged in their outer layers all the say to the present day. By regarding stars as long-term gas preserves, we can still determine the composition of the original "birth gas," even now. By examining stars of various metallicities, trends of the chemical abundances over time can be established. This allows us to trace properties of this early gas and the chemical evolution to the period when the Sun formed

4.6 billion years ago, and even beyond. These details of the past can be acquired only with the help of ancient, metal-poor stars, making them valuable tools in astronomy.

Such information yields important insights into the first and early star formation events and of course the buildup of heavier elements from lighter ones—nucleosynthesis—that were responsible for the production of the first metals. Considering that we humans—with roughly 30 elements constituting our bodies—are the outcome of this billion-year-long production process, it is exciting to be able to trace the origin of the chemical elements this way.

3.3 Element Nucleosynthesis in the Cosmic Kitchen

Astronomy is often regarded as a subfield of physics, in which case it is called astrophysics. There are other research fields bordering on astronomy too, such as chemistry (astrochemistry), biology (astrobiology), and, of course, computer science (computer simulations) and mathematics (statistical analyses). Research on ancient stars in astronomy is closely associated with the chemical elements and their formation. That said, it deals less with the characteristics of the elements themselves than with nuclear chemistry and therefore nuclear physics because every aspect of the physical processes involving nuclei are of considerable interest. Astronomers working specifically with stars and their chemical composition thus become experts in the processes of element formation. With their work, they concentrate on atoms and their characteristics to ultimately learn about their cosmic origins. Hence, this field is called nuclear astrophysics.

From now on, the periodic table of the elements will be a constant companion on our cosmic exploration of the chemical elements. I also have a periodic table in my office. It is a placemat under my keyboard that comes in handy when I quickly have to check something, like, what exactly the atomic number for thulium is. Is it 67? 68? 69? Ah, yes, of course, it is 69! How could I have forgotten that?!

A glance at the periodic table in Figure 3.4 shows how cleverly and informatively it is organized. The name of each element is given in abbreviated form in each little box. The nuclear charge number and atomic

Periodic Table of Elements

1																	2
H																	He
1.008																	4.003

B Solid elements · **H** Gaseous elements · **Hg** Liquid elements (20°C) · **U** Radioactive elements

Atomic number ⇒ Element symbol ⇒ Rel. atomic mass ⇒

1																	2
1 H 1.008																	2 He 4.003
3 Li 6.941	4 Be 9.012											5 B 10.811	6 C 12.011	7 N 14.007	8 O 15.999	9 F 18.998	10 Ne 20.180
11 Na 22.990	12 Mg 24.305											13 Al 26.982	14 Si 28.086	15 P 30.974	16 S 32.066	17 Cl 35.453	18 Ar 39.948
19 K 39.098	20 Ca 40.078	21 Sc 44.956	22 Ti 47.88	23 V 50.942	24 Cr 51.996	25 Mn 54.93	26 Fe 55.847	27 Co 58.933	28 Ni 58.69	29 Cu 63.546	30 Zn 65.39	31 Ga 69.723	32 Ge 72.61	33 As 74.922	34 Se 78.96	35 Br 79.904	36 Kr 83.80
37 Rb 85.468	38 Sr 87.62	39 Y 88.906	40 Zr 91.224	41 Nb 92.906	42 Mo 95.94	43 Tc (98)	44 Ru 101.07	45 Rh 102.906	46 Pd 106.42	47 Ag 107.861	48 Cd 112.411	49 In 114.82	50 Sn 118.72	51 Sb 121.75	52 Te 127.60	53 I 126.905	54 Xe 131.29
55 Cs 132.905	56 Ba 137.327	57 La to 71 Lu	72 Hf 178.49	73 Ta 180.947	74 W 183.85	75 Re 186.207	76 Os 190.2	77 Ir 192.22	78 Pt 195.08	79 Au 196.967	80 Hg 200.59	81 Tl 204.383	82 Pb 207.2	83 Bi 209.980	84 Po (209)	85 At (210)	86 Rn (222)
87 Fr (223)	88 Ra 226.025	89 Ac to 103 Lr	104 Rf (261)	105 Db (262)	106 Sg (263)	107 Bh (262)	108 Hs (265)	109 Mt (266)	110 Ds (281)	111 Rg (272)	112 (272)						

57 La 138.906	58 Ce 140.115	59 Pr 140.908	60 Nd 144.24	61 Pm (145)	62 Sm 150.36	63 Eu 151.965	64 Gd 157.25	65 Tb 158.925	66 Dy 162.50	67 Ho 164.93	68 Er 167.26	69 Tm 168.934	70 Yb 173.04	71 Lu 174.967
89 Ac 227.028	90 Th 232.038	91 Pa 231.036	92 U 238.029	93 Np 237.048	94 Pu (244)	95 Am (243)	96 Cm (247)	97 Bk (247)	98 Cf (251)	99 Es (252)	100 Fm (257)	101 Md (260)	102 No (259)	103 Lr (262)

Figure 3.4. The periodic table of the elements. (*Source*: Peter Palm)

weight are also indicated. The nuclear charge number, also called atomic number, indicates how many protons an atom of a given element has. Every atom is built from three different kinds of particles: protons, neutrons, and electrons. Protons and neutrons make up the atom's nucleus, while the electrons orbit the nuclei in a shell. In order to balance the positive charge of protons, a neutral atom has just as many negatively charged electrons as it does protons.

Some examples of atomic nuclei are illustrated in Figure 3.5. The atomic weight (or atomic mass) indicates the total weight of the protons, neutrons, and electrons. Electrons weigh extremely little compared to protons and neutrons, and hardly contribute to the atomic weight. The hydrogen atom is the lightest atom and has just one proton in its nucleus. A single proton is of course just an ionized hydrogen atom. When an astronomer speaks of an ionized atom, she or he means that the atom has lost one or more of its electrons from its shell. Since hy-

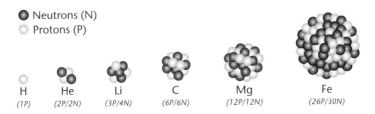

Figure 3.5. Nuclei of various elements. Hydrogen (H), helium (He), and lithium (Li) originated in the Big Bang. Carbon (C), magnesium (Mg), calcium (Ca), and iron (Fe) were later synthesized in stars. (*Source*: Peter Palm)

drogen has just one electron, just the proton is left over in the case of ionization. That electron could have been knocked out by a photon of sufficient energy, for instance.

Helium has two protons, lithium has three protons, and so on. It is the number of protons in a nucleus that decides which kind of element it is. Hydrogen has an atomic number of 1, helium 2, and lithium 3. The periodic table of the elements orders all the chemical elements according to the number of their protons, therefore according to their atomic number. An "element" is just the designation for a nucleus with a very specific number of protons and neutrons. "Light" elements, that is, atoms constituting just a few protons and neutrons, are located in the upper part of the periodic table. The "heavier" elements, richer in protons and neutrons, can be found in the lower part.

In chemistry, the interest is rather on the material properties of the elements and element groups. The elements in one column of the table form a group. All the elements in a group have similar chemical characteristics because their outermost electron shells hold the same number of electrons. The elements arranged next to each other in a row are called periods. They are particularly interesting to nuclear astrophysics and are indicative of the origin and evolution of each element. Elements with similar atomic numbers are generated in similar nucleosynthesis processes in stars and their explosions. Some specific element groups that astronomers repeatedly encounter in their work are listed in Table 3.2.

How does element synthesis proceed in detail? A quick look at the inner structure of a star and its energy production is helpful in this regard. A star is basically a huge spherically shaped object composed of

Table 3.2. Important Element Groups in Nuclear Astrophysics

Element group	Elements and atomic numbers	Location of formation
CNO elements	C (6), N (7), O (8)	Red giant stars
α-elements	Mg (12), Si (14) Ca (20), Ti (22)	Last stages of stellar evolution
Iron-group elements	Sc (21), V (23)–Zn (30)	During stellar evolution and supernova explosions
Neutron-capture elements		s-process in asymptotic giant branch stars and r-process in supernova explosions
—Light	Sr (38)–Sn (50)	
—Heavy	Ba (56)–U (92)	

hot ionized gas called plasma. The Sun is a good example of a typical star for discussing the various processes of element production.

A star has to be in a state of equilibrium to be able to exist for long periods. The force of gravity that pulls the material toward the center and the pressure exerted by hot stellar gas pushing the gas apart have to be in balance. Figure 3.6 illustrates this constant rivalry of the forces. Whenever this balance is upset, the star either collapses upon itself or flies apart.

A star radiates large amounts of energy from its surface. If this lost energy cannot be replaced, a cooling of the gas would result and thus a drop in pressure. The star would collapse. Its energy source is nuclear reactions taking place in its core region. In nuclear fusion, two positively charged nuclei collide with each other at high velocity and form a new, heavier nucleus. Since two positively charged protons repel each other, the electrostatic repulsion has to be overcome first before the pull of the strong nuclear force can attract any particles at close range. Only then can the protons fuse together.

The temperature necessary for nuclear fusion arises at the center of the star through the compression of the gas under its own weight. A temperature of about 10 billion K is needed for hydrogen fusion—only then can the repulsion be overcome. However, the Sun's center is only about 10 million K hot. How can fusion still occur? Thanks to the tunnel

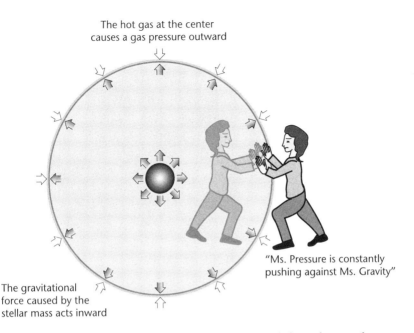

The hot gas at the center causes a gas pressure outward

"Ms. Pressure is constantly pushing against Ms. Gravity"

The gravitational force caused by the stellar mass acts inward

Figure 3.6. A star's outward directed force of pressure has to balance the inward directed force of gravity. Only then does a star remain in equilibrium and neither collapses nor flies apart. (*Source*: Peter Palm)

effect of quantum mechanics, a tiny fraction of all those protons tunnel their way through this repulsion barrier, despite the "cooler" temperature. As a result, enough protons fuse to generate sufficient energy for the Sun to shine.

But where does the energy of nuclear fusion ultimately come from? Einstein's famous formula $E = mc^2$ holds the answer. The nuclear force binds the four particles of a helium nucleus very tightly together. This force is so strong that one helium nucleus is lighter than four individual protons. Compared to those four protons, 0.7% of the mass is "missing." This apparently tiny mass deficiency (also called mass defect) corresponds to an energy according to the relation $E = mc^2$ that is released each time a helium nucleus is formed. This is the Sun's source of energy.

Let us assume that the Sun consists of 100% hydrogen and 10% is fused into helium. Then, 0.7% of the 10% of hydrogen would be converted into energy: that equals about 10^{44} Joules. That amount of energy would be enough for the Sun to shine for roughly 10 billion years. The

Sun is only about 4.6 billion years old. Hence, it has not converted even one-tenth of a percent of its total mass into energy, despite converting 4.2 million tons of matter into radiation every second. In the end, all these processes make it possible for us to watch any star in the night sky.

According to works by Carl Friedrich von Weizsäcker and Hans Bethe from around 1939, there are two ways in which hydrogen can be fused into helium: the so-called proton-proton chain reaction (p-p chain) and the carbon(-oxygen-nitrogen) cycle (CNO cycle). One of the two processes always dominates the hydrogen fusion depending on how hot a star is in its center region and how much energy is needed to counteract the gravitation. Since our Sun is not as hot as other more massive stars in its core, fusion of hydrogen into helium occurs mainly through the p-p chain.

As is depicted in Figure 3.7, in this reaction two protons initially combine into one deuterium nucleus, that is, into heavy hydrogen. In what is called inverse β-decay, one of these two protons spontaneously converts to a neutron, thereby releasing one positron and one neutrino. Table 3.3 describes the various particles involved in nucleosynthesis.

The newly formed deuterium nucleus consists of one proton and one neutron. If such a nucleus collides again with another proton, they fuse into a helium isotope (^{3}He), which emits energy in the form of high-energy photons, as γ-radiation. Said ^{3}He isotope consists of two pro-

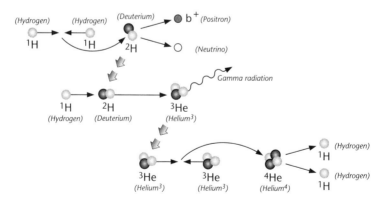

Figure 3.7. In the proton-proton chain, a total of four hydrogen atoms are fused into one helium nucleus in three steps. Energy is released in each step. (*Source*: Peter Palm)

Table 3.3. Overview of Elementary Particles

Particle	Commentary
Proton	Positively charged particle
Neutron	Chargeless particle
Electron	Negatively charged light elementary particle that occurs in the atomic orbital shell
Positron	Positively charged light elementary particle that has the same properties as an electron apart from its charge (the electron's antimatter particle); it is released during radioactive β^+-decay
Neutrino	Virtually massless, chargeless elementary particle that is released during β-decay
Photon	Particle of light

tons and just one neutron. It means that its atomic number, that is, the nuclear charge, is the same as for a regular helium atom (^{4}He) being composed of two protons and two neutrons. Just the mass number of ^{3}He is lower by 1. If two ^{3}He nuclei, in turn, collide with each other, they combine into a single ^{4}He atom. In this nuclear reaction the two leftover hydrogen nuclei are released for them to undergo hydrogen fusion again.

Different from the p-p chain, the CNO cycle requires the presence of at least small amounts of carbon in the star because carbon nuclei are needed as catalysts. This "initial carbon" normally originates directly from the birth gas cloud and is thus evenly distributed throughout the star. To start the CNO cycle depicted in Figure 3.8, temperatures of 30 million K are required in the stellar interior.

In a first reaction, one hydrogen nucleus fuses with a carbon nucleus (^{12}C) into a nitrogen nucleus (^{13}N). As this kind of nitrogen is radioactive, it disintegrates during what is referred to as β^+-decay. One proton transforms into one neutron and two lighter particles, a positron and a neutrino, are ejected. In this decay, the nitrogen nucleus becomes a carbon isotope with mass number 13. It has the same atomic number but a higher mass number than the initial carbon nucleus. If another proton then collides with this carbon isotope, the result is a nitrogen nucleus (^{14}N).

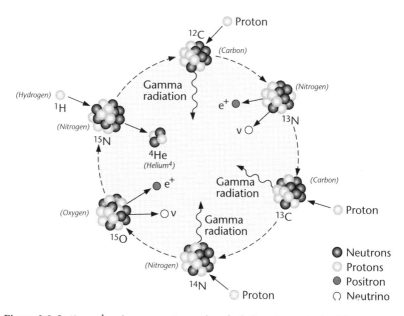

Figure 3.8. In the carbon(-oxygen-nitrogen) cycle, helium is synthesized from hydrogen atoms in a total of six steps starting from the top. (*Source*: Peter Palm)

The outcome of another proton-capture is an oxygen nucleus (^{15}O). This oxygen nucleus, in turn, is radioactive, emits a positron and a neutrino, and thereby transmutes into a nitrogen nucleus of mass number 15. If, finally, one more proton collides with this nitrogen nucleus (^{15}N), a helium nucleus (^{4}He) can be ejected. Helium nuclei are generally called α-particles. In this last process the nitrogen simultaneously transforms into a carbon nucleus of mass number 12, which is the same as the initial carbon nucleus.

The essential difference between the p-p chain and the CNO cycle is that a weak decay, independent of temperature, is occurring at the beginning of the p-p chain. That means that the rate at which the p-p chain generates energy is proportional to a low power of temperature. The CNO cycle energy output, on the other hand, scales with temperature to a much higher power. Due to the small atomic numbers of the elements involved in p-p chain reactions, however, the electrostatic Coulomb repulsion force is much smaller than for those involved in the CNO cycle. Consequently, at lower stellar core temperatures, such as in the Sun, the p-p chain has the advantage of supplying the required energy. For higher temperatures, the CNO cycle, with its steeper temperature dependence, quickly overtakes, becoming the main hydrogen fusion process.

In astronomy, the process of hydrogen fusion is usually called hydrogen burning. It supplies a star with the necessary energy to maintain an equilibrium between the gravitational and pressure forces for 90% of its lifetime. That way, the star can keep its surface temperature, brightness, and luminosity relatively constant while burning its entire hydrogen fuel reserve in its core.

The energy released by hydrogen burning permeates through the whole star from its core upward to the surface, where it is radiated away as light. There are a number of ways for energy to be transported through the star: energy transfer by thermal conduction, radiation, or convection. Conduction of heat does occur in stars but is not particularly efficient. In the process of radiation, photons propagate at the speed of light but have to fight their way through the stellar material. Since photons are frequently scattered, absorbed, and reemitted on their way through the ionized plasma, the stellar interior is quite impenetrable to light. The transfer of energy by radiation is very time-consuming for that reason. Furthermore, there are regions in all stars that are completely impassable to photons. In those layers, convection takes over energy transportation. Energy is carried by rising packets of gas that transport it to the surface.

Anyone can experience the effects of each of these heat-transfer mechanisms in a simple experiment with a burning candle. By putting your finger alongside the flame, you can feel its heat. This is thermal radiation. If you hold a metal pin into the flame from the same distance as your finger was before, you will burn your fingertips on the pin after a few moments. This is thermal conduction. If you position your fingers above the flame the same distance away, your fingers will immediately be singed in the rising hot air. This is convection. This example nicely demonstrates that thermal conduction in gas plays only a very minor role compared to thermal conduction in metal. Moreover, convection transfers heat much more efficiently than thermal radiation.

Hydrogen burning at the center of a star is followed by other burning stages. After all, the star still has 10% of its life left. After the central hydrogen is used up, slowly but surely the hydrogen burning works its way outward in a giant burning shell around the core. This process is also

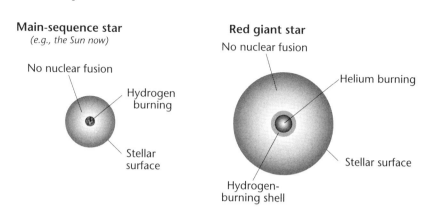

Main-sequence star
(e.g., the Sun now)

No nuclear fusion

Hydrogen burning

Stellar surface

Red giant star
No nuclear fusion

Helium burning

Stellar surface

Hydrogen-burning shell

Figure 3.9. Schematic view of central hydrogen burning (*left*: in a main-sequence star) as well as a later evolutionary stage (*right*: in a giant star) where helium is fused into carbon in the core and hydrogen is fused into helium in a burning shell. (*Source*: Peter Palm)

referred to as shell burning. Figure 3.9 depicts central hydrogen burning in a relatively unevolved star compared to one that has progressed considerably further into the next burning stage.

Since the inner region, now composed of helium, is no longer producing any energy and is just releasing heat, over time the star begins to contract. It continues to heat up as a result. This happens until the center is hot enough for helium to fuse into heavier elements, such as carbon and oxygen from the fusion of three or four helium nuclei, respectively. The fusion of two α-particles, that is, helium nuclei, leads to a beryllium nucleus that by capturing another α-particle is transformed into a carbon nucleus of mass number 12. This is the so-called 3α-process. If such a carbon nucleus captures another α-particle, an oxygen nucleus is finally formed.

Whenever a given burning stage in the center of the star is completed, in other words when its initial element is completely depleted, no energy is produced for a short period of time. Then, gravity gains the upper hand and compresses the star. The density at the center of the star increases, leading to a heating and subsequent ignition of the next stage of nuclear burning. As the elements become increasingly heavy, they need ever hotter conditions for fusion to occur. Carbon burning requires about one billion degrees, which only heavy stars with masses greater than eight times the mass of the Sun can muster. In this burning stage, neon, sodium, and magnesium are produced. Subsequent neon burning continues to synthesize magnesium and oxygen by an energy-dependent separation of one α-particle from neon nuclei. The fusion of two oxygen nuclei leads mainly to the formation of silicon but also to smaller amounts of magnesium, phosphorous, and sulfur. Only if the core temperature continues to rise after oxygen burning has ended can silicon finally be transmuted by the so-called α-process into a range of other elements. The capture of α-particles builds up isotopes of elements with even atomic numbers: magnesium ($Z = 12$) into silicon (14) into sulfur (16) into calcium (20) into titanium (22) into chromium (24) into iron (26) into nickel (28). The heaviest isotope is ^{56}Ni (nickel), with 28 protons and 28 neutrons. It is radioactive and decays via ^{56}Co (cobalt), with 27 protons and 29 neutrons, into ^{56}Fe (iron), with 26 protons and 30 neutrons. Table 3.4 illustrates the properties of the burning phases occurring in a massive star of 20 times the mass of the sun.

In this manner, one element after another is made and then used up until a core of iron and nickel has formed in the stellar center. Up to this

Table 3.4. Nuclear Burning Stages of a Star with 20 Solar Masses

Fuel	Temp. in million K	Density in g/cm^3	Duration in years	Fusion products	Mass in solar masses $M_\odot$
H	37	4.5	8.1 million	He	10
He	190	970	1.2 million	C, O	6
C	870	170,000	980	Ne, Na, Mg	5
Ne	1,600	300,000	0.6	Mg, O	3
O	2,000	6 million	1.3	Si, S	2
Si	3,300	43 million	11.5 days	Fe, Ni	1.5

Source: Data from Woosley, Heger, and Weaver, *Reviews of Modern Physics* 74 (2002): 1015–1073; and Karakas and Lattanzio, *Publications of the Astronomical Society of Australia* 24 (2007): 103–117.

Note: The data in the fourth and last columns depend on the star's mass. The last column specifically indicates the mass of the remaining core synthesized in that fusion stage. For instance, once the hydrogen burning is completed, the star has a helium core weighing 10 solar masses.

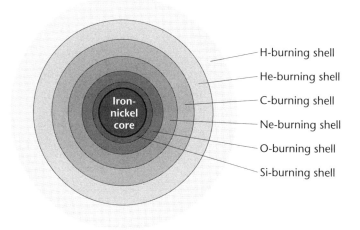

Figure 3.10. Onion-layer model of a massive star with over 8 solar masses at the end of its life. The residual iron-nickel core is surrounded by layers of newly synthesized material. No more energy can be gained from element fusion in the core anymore. (*Source*: Peter Palm)

point, energy could be tapped from fusion, thanks to the mass defect. However, for reasons of nuclear physics, energy is *needed* for the fusion of elements heavier than iron and nickel. This means that instead of gaining energy from this process, a star would have to input energy to form even heavier elements. This does not happen, of course. Instead, the life of a star ends quite abruptly once it has produced an iron-nickel core.

Without another source of energy, gravity wins in the end: The center collapses and an immense shock wave is released that drives the star completely apart in a gigantic supernova explosion. The explosion processes are described in chapter 4. Before the explosion, the star is in its advanced stage of evolution and similar to a giant onion—quite appropriately for our cosmic kitchen. Such an onion layer model of a massive star is illustrated in Figure 3.10. At this time, the shell burning has caused the star to consist of many layers of different elements arranged from the outside in. A hydrogen layer is on the outside, then come layers of helium, carbon, neon, oxygen, and silicon as well as a variety of other elements that had been synthesized in shell burning in smaller amounts. The core of iron and nickel, finally, is at the center.

A shock wave caused by the implosion of the iron-nickel core of a massive star follows its sudden core collapse. When the shock wave hits the compressed core, it bounces back and moves outward through the many element layers of silicon, oxygen, carbon, helium, and hydrogen. A brief but extreme heating of the stellar material occurs as a result. The collapsing iron-nickel core releases so much energy that all the iron and nickel and portions of the surrounding silicon, the products of all that arduous element synthesis throughout the many earlier burning stages in the star's life, are instantly disintegrated into neutrons. These neutrons hit the material in the stellar envelope. Such a bombardment, temporarily coupled with high temperatures such as during a supernova explosion, then allow elements heavier than iron and nickel to be formed, although in much smaller quantities. These so-called neutron-capture elements are found in the lower half of the periodic table. Chapter 5 is dedicated to those elements. The explosion ejects all those newly synthesized elements from the inner layers of the star into outer space, where they mix with interstellar gas.

A star's life, and hence also the course of stellar element synthesis, depends on the star's mass. Stars with an initial mass less than 8 solar masses do not undergo all these burning stages and consequently do not end in a core-collapse supernova. As will be explained in chapter 4, only elements up to oxygen can be synthesized in these lighter stars. Their masses and resulting gravitation are insufficient to produce either high enough core temperatures or the necessary pressure for further stages of burning. Massive stars are exclusively responsible for the production of the heavier elements. Thus, stars of low mass play entirely different roles in the manufacturing of elements and the chemical evolution of the Universe.

A nice illustration of stellar evolution and nucleosynthesis is the star Betelgeuse. It is located in the Orion constellation as the left "shoulder star" of the brave Hunter. Plate 3.A shows the position of Betelgeuse in Orion. The orange-red light emitted by Betelgeuse is easy to see with the naked eye, quite unlike that of the other stars in Orion such as the white-blue right-hand "foot star" Rigel. At some 3,200 K, Betelgeuse is very cool. This explains its reddish appearance. Thus, it falls under the M category of spectral classifications. This roughly 20-solar-mass red supergiant is likely only about 10 million years old and is already in the final stages of its life. This means that hydrogen burning in the core is already over and at the moment it is fusing helium into carbon and oxygen. It will soon—meaning within the next million years—explode as a supernova. Large quantities of matter are already flowing off its surface into space in a strong stellar wind. Once the new carbon and oxygen are transported to the surface, Betelgeuse will immediately release much of this newly synthesized material back into the surrounding gas, even before the star has ended its life.

Rigel, on the other hand, is still in the hydrogen-burning stage. Strictly speaking, Rigel is a binary star, although its companion is considerably fainter. This primary star with 25 solar masses is about one million years old and is classified as a hot B star. It shines with a white-blue color. Both these stars allow us to witness the different stages of stellar evolution and to see element synthesis "in action." We just have to know where to look for them in the sky.

Small, simple, and very inexpensive spyglass telescopes have been available for some years now: the Galileoscope with a 5-cm lens capable of 25- to 50-fold magnification, was developed as part of a scheme by the International Astronomical Union in 2009 in celebration of the International Year of Astronomy. The idea behind it was to offer as many people as possible the opportunity to explore the celestial sky with simple tools. The vast wonders of the cosmos can immediately be experienced with such a miniature telescope, or even just a good pair of binoculars. Not only different stars with their variegated colors, such as Betelgeuse and Rigel, but also—with a little background knowledge—the nebulous regions of star formation, gas clouds, and galaxies can be admired. Suddenly, these distant objects become accessible to us all.

Stars are responsible for the chemical diversity of the elements in the Universe and their abundances in space. Over the course of billions of years, each atom has been synthesized in a star, one by one, again and again. This process has gone on to this very day and will continue into the future. With this knowledge, we can now understand the cosmic origin of the periodic table. Although most of us know the periodic table only from chemistry classes at school, in the end it is astrophysics together with nuclear physics that inform us how all those elements formed inside stars.

3.4 Stellar Diversity

There are plenty of stars in our Galaxy, probably between 200 and 400 billion. That makes an average of about one stellar birth per month since the Big Bang. Elements are continually being synthesized in all those stars, constantly driving chemical evolution forward. The whole population of stars is like a bustling cosmic zoo that includes quite a few exotic species besides any native animals. Although it does not always appear as such, when you are looking at the sky or at pictures of stars, there is an enormous diversity among them. Annie Jump Cannon likely came to this conclusion 100 years ago when she was compiling her spectral classifications. The individual "personality" of each star is certainly written in its spectral signature.

Just as in any group of animals or people, there are larger and smaller stars. Moreover, stars can also alter their sizes. Every star expands to become a so-called red giant toward the end of its life. The giant star Betelgeuse, for instance, has a radius 1,200 times larger than the Sun's because it is nearing its end. This is wider than the radius of Jupiter's orbit around the Sun. Our Sun, by contrast, is still many billions of years away from its final stage, so it is a rather small, unbloated star.

The size of a star does not necessarily imply anything about its mass, though. Stellar mass is an important quantity in astronomy. The Sun with its 2×10^{30} kg (which corresponds to 333,000 times the mass of Earth) serves as a unit of measure for stellar mass: the solar mass. The range of stellar masses is very large. The minimum mass of a star is about 0.1 solar masses, otherwise the star is not heavy and thus hot enough for nuclear fusions to ignite at its center. A star is called a "low-mass star" when it weighs less than 2 to 3 solar masses. Strictly speaking, stars between 2 to 3 and 8 solar masses are neither low-mass stars nor massive. We call them "intermediate-mass stars." A star is considered massive if it is 8 or more times heavier than the Sun. Some stars have 20 solar masses or even more, and a handful are even considerably more massive than that, having up to 100 solar masses. Number-wise, there are many, many more low-mass stars than massive ones. For every massive star there are about 1,000 intermediate-mass ones and 10,000 low-mass stars. Actually, most stars have masses considerably less than one solar mass, such as 0.3 solar masses or even less.

As the mass increases, so does a star's rotation. Less massive stars rotate slowly as opposed to more massive ones with over 5 solar masses. However, all stars will rotate slower as they age, particularly the long-lived low-mass stars.

Stars are also differently colored because they have different surface temperatures. The same effect is nicely observable when using a charcoal barbeque grill. Normally the charcoal glows a faint dark red. After blowing some fresh air onto it, the burning is rekindled, causing the burning temperature to rise and the charcoal glows first with a bright red, then orange, then yellow-white and perhaps even bluish. This is the color palette of stars.

Some stars are located closer to the Sun than others, depending on their positions in the Milky Way. To the eye, stars of the same luminosity at different distances from us seem to differ in brightness. Farther-away street lamps also seem fainter than nearer ones. A very luminous but distant star, however, can certainly appear to be brighter than a fainter nearby object. For this reason, the apparent brightness of an object in the sky does not reveal much about its actual luminosity.

The luminosity of most stars changes only very slowly over the course of their lives, and when it does it is during the star's final stages of evolution. Hence, such changes cannot be observed directly. Nevertheless, many "variable stars" do exist. Due to their regular pulsations, their luminosity and thus also apparent brightness vary significantly over short periods, ranging from minutes to many hundreds of days. These variations are observable. The class of so-called δ (delta) Cephei stars—named after their prototypical representative—is worth mentioning here. They are very bright giant stars with magnitudes fluctuating in clearly measurable periods of a few days. As a result, they often serve as a galaxy's "lighthouses." Their enormous luminosities make them even observable in other very distant galaxies—Edwin Hubble discovered this already around 1920.

The next closest neighbor to our Sun is the small, very cool dwarf star Proxima Centauri in the Centaurus constellation. It is just 4.2 light-years away from the Sun. Right next door to it is the Alpha Centauri system. It is a binary system composed of two Sunlike stars. They are about 4.3 light-years away from us. Of all stars, 60% are born in binary-star systems—consequently, systems such as Alpha Centauri are not rare.

Double- or multiple-star systems contain two or more stars or one star and a compact object, such as a neutron star. White dwarfs, neutron stars, and black holes are more exotic versions of "stars," as no nuclear fusion occurs inside them. Chapter 4 covers those objects in greater detail, as, strictly speaking, they are not really stars. They are merely the compact remains of stars of different masses.

Binary-star systems are bound by their mutual gravitational attraction making the objects revolve around each other. If the stars are far apart, they can peacefully travel along their orbits over billions of years. If, however, they are orbiting their mutual center of mass along a close-in orbit, gravitational interactions can occur between them. There is often an exchange of mass between such stellar pairs, by which the more massive primary star donates a portion of its outer envelope to the less massive partner. This exchange of material between two stars could be regarded as a cosmic love story, and further details about those processes are presented in chapter 5.

The spectral class of carbon stars, found by Angelo Secci prior to 1900, can be attributed to such a mass transfer onto the lower mass binary-star companion. Every 10,000 years, explosive helium burning can occur in the outer layers of the evolved primary star. It is called a "helium shell flash." The carbon formed in the process (along with other elements) is transported to the stellar surface by convection. The primary star then turns into a "classical" carbon star. If the less massive star is orbiting such a giant star at close range, however, significant mass transfer can occur following the flash. In that case, parts of the primary star's outer atmosphere containing all that carbon is transferred to the companion, which also changes its surface abundance and becomes a nonclassical carbon star.

The stars we see in the sky are mostly individual stars or stars in a binary-star or multiple-star system. But then there are also many stars that occur in star clusters. For example, the famous Pleiades is an open star cluster actually comprising about 500 young stars that formed about 115 million years ago. It is shown in Plate 3.A. Between five and ten of the brightest stars are visible to the naked eye in the Taurus constellation and make up the famous "Seven Sisters." Aside from such younger open clusters there are also the so-called globular clusters, often consisting of a million stars. These immense structures are generally very old, about 10 to 12 billion years, and are located on the outskirts of the Milky Way. Even though most of the individual stars are not particularly luminous, globular star clusters still appear relatively bright in the sky, due to the sheer number of their members. Many of these clusters are observable with a small amateur telescope and look like fuzzy spheres.

Naturally, there is a huge variety of younger and older stars in the Universe. Ongoing star formation in dense gas clouds constantly produces newborn stars. The Sun, at 4.6 billion years, is still relatively young in this regard. The oldest stars, by contrast, are about 13 billion years old and therefore almost as old as the Universe itself.

For completeness, it should be reiterated here that the chemical composition of stars is, of course, an important distinguishing feature. The great majority of all stars have compositions similar to that of our metal-rich Sun. Relatively few stars have a low metallicity—the lower the metal abundance, the rarer the star because it comes from the early era of the Universe more than 10 billion years ago.

These diverse metallicities influence the color of stars in their own way. Metal-poor stars appear a little bit bluer than metal-rich stars. Metals, such as iron, absorb particularly at short wavelengths, that is, blue light. Metal-poor stars thus characteristically absorb this kind of light less compared to metal-rich stars. As a result, they emit relatively more blue than red light. Furthermore, their low concentrations of metals reduce the scattering of photons in their atmospheres, allowing us to look deeper "inside" the stars. Those stars thus also look hotter and more luminous than metal-rich stars and their light is again somewhat bluer.

Finally, a few dozen stars are known to have characteristics extremely similar to those of the Sun. They are referred to as solar twins. The Sun is the home star to the entire Solar System with different kinds of planets. Other stars surrounded by their own planets are already known but so far none of those planets closely resemble Earth. Much progress is being made in this field right now, so it is probably only a matter of time before Earthlike planets are found. Such discoveries do not immediately solve the question about life in space, though. Our knowledge about the details of the beginning of life, planet formation, and the host star's role is still not detailed enough.

Hidden behind those tiny twinkling lights in the sky is thus an incredible range of all sorts of different stars. Their diversity offers astronomers countless opportunities to better understand and explore the cosmos.

CHAPTER 4

STELLAR EVOLUTION—FROM BIRTH TO DEATH

On a clear night a few thousand stars are visible to the naked eye. The celestial sky is thus like a photograph of a big crowd of people. We are dealing with a cross-section of an entire population and much can be learned about the characteristics of humans—or stars—from such an image. Just like we people are born as babies, grow up as children and teenagers, and later die as old and experienced adults, all stars go through a cosmic life cycle of their own consisting of a number of stages.

By observing different kinds of stars, it is possible to gain clues about their nature and this life cycle. It becomes evident that the different stages of stellar evolution are closely connected to the various processes of nuclear fusion occurring in their inner regions. Keeping this in mind, we can now examine the evolution of stars in greater detail, from their births to their gigantic supernova explosions.

4.1 Sorting Stars

At first glance, a crowd of people looks messy and unstructured. But with a few little tricks the apparent jumble can be sorted to gain some order. For example, all the participants can be grouped by the color of their shirts. Or else, a group of people can be divided by body height, age, or gender. Important conclusions about the group as a whole can be drawn from this information. Sorting stars is very similar. Surface temperature, luminosity, and chemical composition are the characteristic features. With knowledge of these three properties, it is possible to classify stars because they describe basic details about their nature.

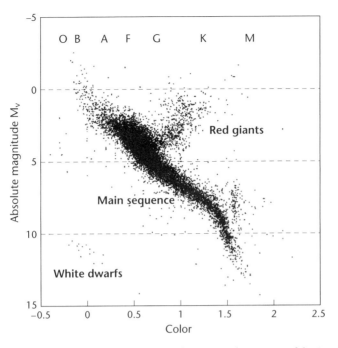

Figure 4.1. The Hertzsprung-Russell diagram for stars in the vicinity of the Sun. The different branches and regions are easily recognizable. (*Source*: Peter Palm; data from the HIPPARCOS and TYCHO catalogues, ESA, 1997, ESA SP-1200)

But how do astronomers sort stars? The Hertzsprung-Russell diagram, named after the Danish astronomer Ejnar Hertzsprung and the American astronomer Henry Norris Russell, provides the necessary means to bring order to the diversity of stars. It is one of the most important tools in stellar astronomy and is depicted in Figure 4.1. Since the various types of stars differ in surface temperature, luminosity, and composition in a very specific way, stars cluster into different and fairly obvious branches and sequences in this diagram.

In the Hertzsprung-Russell diagram, the surface temperatures of stars are plotted along the horizontal axis and their luminosities along the vertical axis. Hot stars are located on the left side and cool ones on the right side. Less luminous stars are in the lower part, while more luminous ones are in the upper part. The surface temperatures and luminosities are either calculated theoretically or are based on observational data. There are a few equivalent physical quantities that describe

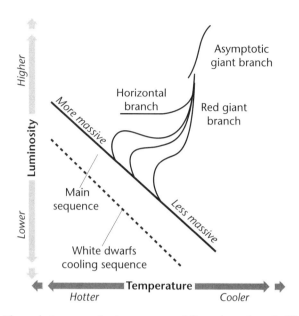

Figure 4.2. The evolutionary paths that every star follows throughout its life, depending on its mass, rendered in a schematic Hertzsprung-Russell diagram. (*Source*: Peter Palm)

temperature as well as luminosity. Examples include a star's color or spectral class, which indicate its temperature and absolute brightness (magnitude), a measure of luminosity. The gravity at a star's surface ("surface gravity") is also a measure of its luminosity.

In the Hertzsprung-Russell diagram, all stars fall into place in different sequences and branches that have been assigned different names and are schematically illustrated in Figure 4.2. The so-called main sequence is easiest to recognize. It is clearly visible running from the bottom right-hand corner of the diagram to the top left.

The point on the main sequence at which stars "bend" in the direction of the "red giant branch" is called the "turn-off point." The giant branch is then clearly visible in the cooler part of the Hertzsprung-Russell diagram: it runs diagonally upward toward the right, like the branch of a tree.

Around 1910, when the Hertzsprung-Russell diagram was developed, it served merely as a classification scheme for stars. The physical context was completely unknown and unexpected at the time. The peculiar grouping of the stars did invite speculations (that were generally wrong at first) about a temporal evolution of stars. Still, a basic theoretical understanding of the position of the stars in the Hertzsprung-Russell diagram could finally be gained with the knowledge that stars draw their energy from nuclear fusion and that those processes drive stellar evolution. What we know today about the physical conditions of these different stages in a star's lifetime largely relies on complex computer-based model calculations of stellar structure and evolution, a tried-and-true subfield of astrophysics. The results of those calculations are compared against observed characteristics of real stars, for instance, their locations in the diagram, and are gradually improved and optimized.

It is remarkable that with the help of this diagram we can catch a glimpse of what is going on inside the core of each star and in which evolutionary stage the observed star happens to be. Its position in the Hertzsprung-Russell diagram sensitively depends on which elements are undergoing fusion in the core region and which of the burning shells are layered above it. Consequently, a star and its surface react to these fusion processes, for instance, with an expansion of the envelope in response to a loss of fusion energy at the core. We are able to observe only the consequences of nuclear fusion since observations document what is occurring at a star's surface but never in its interior. The only exception is helioseismology, which records pressure waves from the core region of a star.

Astronomers today can easily read off what a particular location in the Hertzsprung-Russell diagram means for the star and its evolutionary status. But exactly how does this evolution proceed? Every time nuclear fusion processes change in the core, the star enters a new phase of its life. Sections 4.3 and 4.4 describe those stages in detail for stars of different masses taking into account element nucleosynthesis and the positions in the Hertzsprung-Russell diagram. Here, this evolution is briefly summarized to provide a general overview.

Initially, a star takes its place on the main sequence of the diagram once nuclear fusion has ignited. Only then does it have a stable core temperature and luminosity and is actually a proper star. Such stars are duly called main-sequence stars. Where exactly a star is located on the main sequence is governed by its mass. There is a close relationship between mass and luminosity for main-sequence stars. The bottom right

Table 4.1. Lifetimes on the Main Sequence and Total Lifetimes of Stars with Different Masses

Initial mass in solar masses	Time on the main sequence in millions of years	Total lifetimes in millions of years
0.8	2.0×10^4	3.2×10^4
1	9.2×10^3	1.2×10^4
2	8.7×10^2	1.2×10^3
5	78	102
15	11	13
25	6.7	7.5

Source: Data from Woosley, Heger, and Weaver, *Reviews of Modern Physics* 74 (2002): 1015–1073.

of the sequence is occupied by low-mass stars with low surface temperatures and low luminosities. The upper left is the region of massive stars with very high surface temperatures and high luminosities. The massive stars then continue their evolution further up in the diagram than the lower mass stars. Our Sun is an example of a low-mass main-sequence star, located about one-third from the lower end of the main sequence.

The time on the main sequence is the longest stage in the life of a star. All stars spend about 90% of their lifetimes there. Table 4.1 lists the lifetimes of stars of different masses on the main sequence.

Since stars stay on the main sequence for so long, it follows that 90% of the observed objects in a random sample are main-sequence stars. A star hardly changes during this long stage, as it is simply fusing hydrogen to helium in its core. Consequently, its position on the main sequence does not change either. The Sun, with a life expectancy of about 10 billion years, has already been on the main sequence for 4.6 billion years. It will reach the next stage only in about 4 to 5 billion years. Less massive stars than the Sun spend a substantially longer time on the main sequence because they live a lot longer. Their much lower luminosities cause them to use up the hydrogen very sparingly. More massive and more luminous stars, in contrast, burn their hydrogen fuel much more rapidly. Accordingly, they reach the end of their lives considerably sooner. The different timescales of the evolution of stars of different masses are illustrated in Figure 4.3.

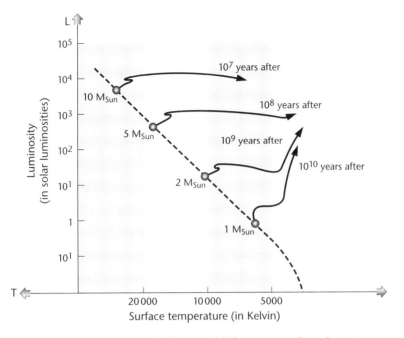

Figure 4.3. Evolutionary time scales for stars of different masses from the main sequence to shortly before the end of their lives. More massive, more luminous stars are located at the upper part of the main sequence, from where they start their evolution. Less massive, fainter stars, such as the Sun, are found further down the main-sequence. (*Source*: Peter Palm)

After the main-sequence stage, a star begins to go through the so-called red giant branch phase. The star leaves the main sequence at the turn-off point to then spend most of the rest of its life on the red giant branch. A star with one solar mass will be a red giant for roughly 1 billion years, whereas for a star with 10 solar masses, this stage lasts only about 1 million years. Strictly speaking, red giants are not really red in color but rather orange. Prominent examples visible to the naked eye are Aldebaran, the "red" eye in the Taurus constellation, and Betelgeuse, the left shoulder of Orion that can be seen in Plate 3.A. The Sun will leave the main sequence and become a red giant in about 4.5 billion years.

Life after the red giant branch as well as the stars' ultimate fate depend entirely on their mass. After the red giant branch stage are two more branches, the horizontal giant branch, or horizontal branch for short, and the asymptotic giant branch. After completing the red giant branch phase, the star jumps from the tip of the red giant branch to the horizontal branch, which is situated roughly in the middle of the Hertzsprung-Russell diagram. From this point, the star will burn helium to carbon in its core. As its name reflects, the horizontal branch is a stage in which a star's luminosity changes very little, but its surface temperature increases steadily. From there, the star moves on to the asymptotic giant branch, which leads to the upper-right corner of the diagram. There, the star awaits the end of its life since nuclear fusion will soon cease. The remnants of low-mass stars, white dwarfs, then move from there into the lower-left part of the diagram before slowly cooling down.

A star cluster's Hertzsprung-Russell diagram is particularly interesting because the distribution of the different stars can be seen particularly well. Since all members of a star cluster are of the same age and have the same composition, their individual masses, and hence their luminosities, are what distinguish them. Different masses lead to different timescales on which each star evolves. The most massive stars have the shortest life spans and sit far up on the main sequence. Owing to their extremely high rate of burning their core hydrogen, they are also the first ones to leave and become red giants. Less massive stars with longer life spans then follow. They are located further and further down on the main sequence.

If you observe a star cluster and then create a Hertzsprung-Russell diagram of all its member stars, information of the cluster's age can be obtained. Depending on the age, you will see only the least massive stars left on the main sequence, because upward of a particular mass, all stars are already red giants or have even reached the end of their lives. Chapter 6 describes star clusters and their Hertzsprung-Russell diagrams in more detail.

In closing, it should be mentioned that for our work with metal-poor stars, we can determine not only the surface temperature but also the surface gravity from the observed stellar spectra. This method is described further in chapter 7. Depending on the kinds of stars, other methods also yield these properties. But no matter how they are obtained, these two physical quantities are suitable for entering a star into the Hertzsprung-Russell diagram since the surface gravity is related to the luminosity via the size of the star. Then it becomes easy to determine a star's evolutionary stage. Especially for determining the stellar chemical composition and subsequent interpretation, this is of great importance.

4.2 A Protostar Forms

Perhaps the prettiest of all constellations—Orion, the hunter—can be seen moving across the night sky on a clear winter night in the Northern Hemisphere. Seen through a small telescope, a splendid little nebulous region appears below the belt stars: the Orion nebula. It is part of an immense molecular cloud consisting of hydrogen, helium, and a small amount of metals. This shimmering pink-purple nebula is 1,600 light-years away and has been photographed countless times, including by the Hubble Space Telescope. Its complex structure of gas and dust looks very graceful and delicate. Plate 4.A shows one of those photographs. Best of all, this nebula is even visible to the naked eye. You only have to know where to look for it: the nebula itself is the middle of the three faint sword stars. Under good conditions, it is easy to spot by eye or with binoculars.

Anyone who seeks out the Orion nebula in the sky can take a look at a cosmic delivery room. This nebula, extending roughly 20 light-years across, is one of the most prominent regions in which active star formation can be observed. The cloud of cool and dense molecular hydrogen and dust in which new stars are being born is impenetrable to visible light, however. It is thus not possible for us to directly witness a "stellar birth." Only with observations in the infrared wavelength region can the various processes in the inner parts of the cloud be investigated.

Many similar nebulae are found in the Milky Way and other galaxies and allow us to study how stars are born. These molecular clouds cool down as their molecules release energy in the form of radio waves, for example. In addition, the surrounding stars' radiation heat the clouds from the outside. The result is turbulence within the cloud, causing it to heat up to a temperature of about 100 K (−173 °C). Individual dense clumps, called globules, are also present in the cloud. They are roughly 10 K (−263 °C) and are shielded from the irradiation by other stars by layers of gas and interstellar dust. This allows them to maintain their low temperature.

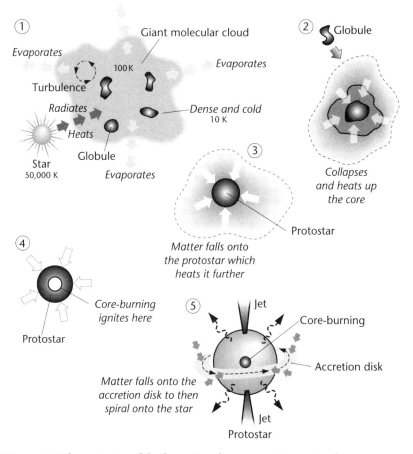

Figure 4.4. Schematic view of the formation of a protostar. In a series of processes, a gas cloud becomes a star with hydrogen burning taking place at its core. (*Source*: Peter Palm)

When the balance between gravity pulling inward and thermal radiation pushing outward is upset in these globules, they collapse and become denser. This and other evolutionary stages are schematically depicted in Figure 4.4. The compression of the gas releases gravitational potential energy, which is converted into thermal energy—heat—and is radiated away into the vicinity. If the gas continues to condense even further, a protostar can finally form.

Hot protostars are giant spheres of gas still inside their dense molecular birth cloud. Although nuclear fusion is not yet occurring, they are already radiating energy into the cloud. This loss of energy is balanced by the release of gravitational potential energy stemming from the contraction of the forming star's gas mass. Despite this radiation loss, protostars continue to heat up.

During this process, new material in the form of gas continues to fall onto the rotating protostar. It does not plunge directly into the star but collects in a disk spiraling around the nascent object. The material is eventually flung off the inner edge of this accretion disk onto the star. The continued accumulation of mass leads to increased weight and pressure of the core region causing it to further heat up.

Like a whale ejecting a fountain of water out of its blowhole as it breathes, a young star also immediately spouts part of the incoming material back into its natal cloud. These fountains, usually referred to as jets, align with the protostar's rotational axis. Where the jet hits the surrounding gas cloud with full force, astronomers detect a bright glow. This is also known as a Herbig-Haro object. As time passes, the increasing radiation and the effect of the jets will gradually free the protostar from its molecular-cloud cocoon.

These young so-called T Tauri stars have immense surfaces and are sometimes up to 100 times larger than the Sun. Correspondingly, the energy they radiate per unit of time as well as their luminosity are rather large. The centers of these forming stars are still too cool for nuclear fusion to ignite, but their steady collapse generates enough thermal energy for them to shine with a fairly steady luminosity for many millions of years. These wild teenager stars still rotate pretty rapidly, about 10 times as fast as the Sun. They also expel a lot of gas from their surfaces which periodically alter their luminosities.

Only with time do these stars get hot enough for their first few nuclear reactions to eventually take place in their core region. At a central temperature of about 1 million K, the already present deuterium nuclei fuse into helium. Deuterium burning begins. It slowly eats its way through the star, steadily raising its temperature before hydrogen burning can finally commence at 10 million K. From that point forward, fusion energy is released and remains the star's source of energy for the rest of its life. With the ignition of nuclear fusion after 100 million years, a young T Tauri star thus becomes a proper adult star. In accordance with its mass, it has then reached its due position on the main sequence in the Hertzsprung-Russell diagram.

4.3 The Evolution of a Low-Mass Star

Throughout their lives, low-mass and intermediate-mass stars of less than 8 solar masses go through very similar evolutionary stages. For this reason, they can be considered together. Once such a star has reached the main sequence, it sustains the fusion of hydrogen into helium in its interior for 90% of its lifetime. Since the temperatures are by far highest at the stellar core, hydrogen burning occurs there and is exhausted there first. What remains is a burning layer above the core where hydrogen continues fusing into helium. This is called shell burning. Then, the newly formed inert core region underneath the hydrogen-burning shell begins to contract and slowly heat up in the process. The excess heat eventually causes the hydrogen burning in the shell to flare up. The layers above the burning shell then expand very rapidly and begin to cool down. Within a very short period, the star moves away from the main sequence to the lower end of the red giant branch.

On the giant branch, the inert helium core of the star continues to slowly contract and heat up. At the same time, the hydrogen-burning shell becomes thinner while further expanding, resulting in an increase in surface area and thus luminosity. As a result, the surface temperature is decreased by several thousand degrees. The extreme expansion also reduces the surface gravity of the star.

Halfway up the red giant branch, elements from the star's interior are transported to the outer surface layers for the first time by immense convective currents circling in the stellar envelope. The CNO cycle that had operated in the core caused some carbon to be converted to nitrogen. Consequently, any dredged up gas is somewhat carbon-poor and nitrogen-rich. The resulting changes of the surface carbon and nitrogen abundances are observable and telltale signs of the CNO cycle operating in the interior.

At the tip of the red giant branch, the helium in the center has become dense and, at 100 million K, also hot enough for the helium core to ignite. Helium then fuses into carbon and oxygen. In low-mass stars, this ignition causes the so-called helium flash. The new and sudden source of energy heats up the center enormously and catapults the star into a new position in the Hertzsprung-Russell diagram. The star has entered a new phase in its life. Compared to hydrogen burning, which in the Sun's case lasts almost 10 billion years, helium burning finishes within just about 100 million years, hence, 100 times faster.

Helium burning provides a renewed energy source, which leads to a rapid decrease in the luminosity of the hydrogen burning shell and the star's total luminosity. The star contracts, thus gaining back a higher surface temperature. The star is finally located on the horizontal branch. The ignition of helium burning is less dramatic in intermediate-mass stars. The star does not jump onto the horizontal branch but evolves in that direction. A star's structure on the horizontal branch resembles that of the main sequence. The only difference is that now the star burns helium into carbon and oxygen in its core while hydrogen is still fusing into helium in the burning shell above it.

The horizontal branch is intersected by the so-called classical instability strip, a region in the Hertzsprung-Russell diagram containing variable and pulsating stars. Stars burning helium into carbon and oxygen in their interiors that are positioned on the horizontal branch are generally pulsating during this phase. They follow the period-luminosity relation, making them important for astronomical distance determinations.

After the star has entirely burned the helium in its center, similar changes take place as they did at the end of core hydrogen burning. A helium-burning shell forms at the outer edge of the new carbon-oxygen core and the star moves from the horizontal branch to the base of the asymptotic giant branch.

As time passes, the helium-burning shell gets thinner and thinner and becomes thermally unstable for that reason. It reacts very quickly to any addition or removal of heat. Any rise in temperature, no matter how small, as caused by the burning core below, immediately leads to an uncontrolled rise in temperature in the helium-burning shell. As the 3α-process is highly temperature dependent, the carbon production rate shoots up, in turn causing the temperature to rise even more. The gas in and above the helium-burning shell expands as a result.

Thermal instabilities cause the temperature in the thin helium-burning shell to continue to rise despite the expansion. At the same time, the surface temperature decreases due to the expansion of the layers above the helium-burning shell. This includes regions where the hydrogen burning shell is still operating. Consequently, the luminosity of the hydrogen-burning shell drops drastically. After only a few dozen years, however, the helium-burning shell has expanded enough for it to regain thermal stability. Thus, any further expansion leads, again, to cooling and not to additional heating.

The overall contraction of the stellar envelope causes the temperature at the location of the hydrogen-burning shell to rise, thus increasing the fusion rate in the shell. "Thermal pulses" originating from the bottom of the helium-burning shell occur at fairly regular intervals of a few thousand years in all stars with both helium and hydrogen burning shells.

Before such a thermal pulse occurs, the star's luminosity is dominated by the luminosity of the hydrogen-burning shell. During the pulse, the luminosity of the helium-burning shell becomes much brighter than the hydrogen-burning shell for a few decades. Most of this energy goes into expanding the stellar envelope, though. In the end, the star's total luminosity decreases despite the thermal pulse. During the pulse, a short-lived convection zone forms above the helium-burning shell that reaches almost up to hydrogen-burning shell. Material from between the two shells can thus be mixed into the helium-burning shell. Simultaneously, other nucleosynthesis products are flushed upward to just below the hydrogen-burning shell.

After the pulse is over, the hydrogen-burning shell sinks back down again, right into the region that was mixed during the pulse phase. The star's regular convection zone follows and sinks down to just above the new position of the hydrogen-burning shell. This new position is actually further inward than the position of the hydrogen-burning shell during the pulse. In this way, the hydrogen-burning shell "collects" the products of the helium-burning shell that had been mixed outward. From there, the convection zone carries the mixed material to the surface, just like a dumbwaiter would transport food to the floor up.

The helium-burning shell fuses helium (^{4}He) into carbon (^{12}C) and oxygen (^{16}O), and the hydrogen-burning shell then converts this carbon and oxygen into nitrogen (^{14}N). The nitrogen, however, remains underneath the hydrogen-burning shell even as the shell burns further outward between any two pulses.

In the pulse that follows, the nitrogen is mixed down into the helium-burning shell by the short-lived convection zone operating between the shells. Nitrogen is immediately burned in a chain reaction that produces fluorine (^{18}F), then oxygen (^{18}O), and then neon (^{22}Ne). In intermediate-mass stars, fusion of neon into magnesium (^{25}Mg) follows. The neutrons released in this chain of reactions form the neutron source for the s-process described in section 5.1.

For low- and intermediate-mass stars, this is the end of their evolution. They do not have enough mass to cause a contraction and sufficient heating of the carbon-oxygen core to ignite carbon fusion. The remaining sources of energy are the helium- and hydrogen-burning shells. The closer these two burning shells approach the stellar surface, the stronger the consequences of the thermal pulses become: The star begins to eject its outermost layers piece by piece until, during the final pulse, the entire material above the helium-burning shell is suddenly blasted into interstellar space. What is left of the star is its dead, still extremely hot carbon-oxygen core—a white dwarf. It is surrounded by a colorful, glowing gaseous cloud, a planetary nebula.

Planetary nebula? What a peculiar name! Planets have actually nothing to do with those nebulae! To the observers of the 18th and 19th centuries, many of those nebulous spots looked at first like the faint disks of the distant planets Uranus and Neptune through their telescopes. As can be seen in Plate 4.B, planetary nebulae display a marvelous array of colors and shapes owing to the white dwarfs at the center exciting their surrounding gas.

During the process of shell ejection, the surface temperature of the star rises up to 100,000 K because the lost outer layers of the star expose hotter layers below. In the end, the mass of the completely exposed stellar core, nearly independent of the initial mass of the star, is between one-half and one solar mass. Also, white dwarfs do not sustain any nuclear fusion anymore. This remnant depicts the last stage in the star's life. Within a short period, this stellar core moves from the upper end of the asymptotic giant branch horizontally to the left in the Hertzsprung-Russell diagram. This is simply due to the rising surface temperature of the exposed core. From there, it finally jumps into the white dwarf region, which is located toward the bottom left, below the main sequence. This region is a kind of cosmic graveyard since those white dwarf remnants remain there for many billions of years while slowly cooling down until they are as cool as outer space with 2.7 K. They are too cold to be observable anymore long before that, though, as this cooling process happens very slowly. A Sunlike star requires between 10 and 15 billion years to change from a main-sequence star into a white dwarf. Another 7 billion years are needed for it to cool down completely. Since the Universe is just 14 billion years old, only few cold white dwarfs exist yet.

Most white dwarfs are composed of carbon and oxygen. Since contracting after the helium burning has ended, the progenitor star to each white dwarf has an extremely dense carbon and oxygen core. In fact, these cores are so dense that they cannot collapse any further. This means that the extremely compact white dwarf has a mean density of about one billion kg/m^3. The stellar matter is "degenerate" at such a high density. Quantum physics prevents matter composed of protons, neutrons, and electrons from being packed more densely. The entire mass of roughly one solar mass has been compressed into a volume that corresponds to Earth's, when taken at a diameter of 10,000 km. If, however, a white dwarf exceeds a mass of 1.4 solar masses, for example through the accretion of additional matter from a binary companion, the degenerate object becomes unstable. As will be described in section 4.5, the white dwarf cannot continue to exist as such and explodes as a supernova.

For completeness, it should be mentioned that stars of masses greater than 8 solar masses can also end up as white dwarfs. This is the case for stars with particularly strong and long-lasting stellar winds, and also for stars that are regularly subjected to instabilities. As a consequence, they all lose a significant portion of their mass to the interstellar medium already during their main-sequence and red-giant-branch phases. If their total mass is decreased to below 8 solar masses at the end of their lives they likely follow the evolutionary path of intermediate-mass stars.

4.4 The Evolution of a Massive Star

Although more massive stars have substantially larger stores of hydrogen than lower mass stars, they do not actually live any longer. Quite the contrary. Considering that a star's luminosity increases by the third power of its mass, those behemoths shine extremely brightly. The fusion reactor in the interior of a massive star runs at an incredibly higher pace than our low-mass Sun's, for instance. Its fuel consumption is proportionately higher. The price that massive stars pay for their voracious appetite for energy is a shorter life span. The nuclear fuel of a star with 20 solar masses is already spent after just a few million years. By contrast, the core-burning material available to a very low-mass star with just one-tenth of a solar mass lasts for over a trillion years.

Massive stars also spend 90% of their lives on the main sequence, where they burn hydrogen into helium. Just as in low and intermediate mass stars, the newly formed helium core contracts while hydrogen shell burning moves outward, layer by layer. The star moves onto the red giant branch and soon afterward helium burning begins in its core. The subsequent stages for massive stars proceed differently, however— namely, more rapidly and extreme, and additional nuclear fusion stages occur.

Helium burning takes place during the asymptotic giant branch phase. Massive stars form a carbon core large enough for temperatures of over 1 billion K to be reached during the contraction phase subsequent to helium burning.

At such high temperatures and densities, large amounts of neutrinos form in the stellar interior. Since neutrinos hardly interact with matter, they leave the star at the speed of light without contributing any counteracting pressure to gravity in the way photons would. Hence, the energy invested by nuclear fusion to produce neutrinos is immediately lost to the star and adds nothing to its stability. The stellar core continues to contract and heat up, causing the nuclear reactions to proceed increasingly rapidly. The hotter the core region, however, the more neutrinos are generated. Nuclear burning in the star accelerates. That is why carbon burning in a star with 20 solar masses only lasts 1000 years. Without the energy loss to neutrinos, it would have taken 10,000 years.

Examples of individual burning stages and their timescales for a massive star are listed in Table 3.4.

More stages of nuclear burning follow that then occur at even higher temperatures in the stellar core. Neutrino losses increase accordingly and beyond measure. The final stage, silicon burning, occurs at about 3 billion K, and the star radiates as much energy in the form of neutrinos within a single second as the Sun does in the form of light in one million years. The silicon burning is completed in just 10 days, whereby more than one solar mass of silicon is converted into iron. At this point, the star is composed of an iron and nickel core from which no more energy can be gained by fusion. Many shells containing the synthesized elements from the previous burning stages are layered above this core. The star has become a giant "element onion," as can be seen in Figure 3.10.

Particularly the advanced burning stages in the core of a massive star proceed in a very short time. However, the outer envelope of the star is not affected by the processes in the core because the envelope cannot react that quickly to any changes. The star just continues to move along the asymptotic giant branch in the Hertzsprung-Russell diagram.

Then, at some point, conditions become extreme. The star explodes as a gigantic supernova. Since the Hertzsprung-Russell diagram refers only to physical descriptions of the external stellar surfaces, this event cannot be "read off" from the diagram. The course of such an explosion is described in detail in section 4.5. The star is not entirely torn apart, though. The remnant at the end of a low- or intermediate-mass star's lifetime is a white dwarf. In contrast, a massive star leaves behind a so-called neutron star or black hole.

During the supernova explosion, the iron-nickel core with a mass between 1.4 and 3 solar masses very rapidly collapses under its own gravity. Any initially still free electrons are pressed inside their nuclei, whereby the protons are transformed into neutrons in what is called inverse β-decay. The quantum mechanical degeneracy pressure of neutrons created in this process halts the collapse of the object at a radius of 10 to 20 km. A stable neutron star has formed. Similar to a white dwarf, it crystallizes and cools over the course of eons.

It is difficult to observe these barely glowing objects. Many of them can still be detected, though. When a neutron star forms from a slowly rotating massive star, the same thing happens as when an ice skater begins to spin. The rotation accelerates when the arms are drawn in. Since the neutron star is a collapsed object, it rotates very rapidly. These neutron stars emit narrow beams of radio waves and gamma radiation from their magnetic poles. When such a beam hits Earth, we see the object periodically light up once every rotation, just like a lighthouse beam. These neutron stars are called pulsars and were an accidental discovery in 1967 by the British astronomers Jocelyn Bell-Burnell and Anthony Hewish.

If the original mass of the star is greater than 20 solar masses, the final compact remnant at the end of the star's evolution would be too massive to become a neutron star. Neutron stars can form only from initial iron-nickel cores of less than 3 solar masses. Only then is the degeneracy pressure exerted by the neutrons able to stop the gravitational collapse to produce a compact object. What happens to the star if it cannot end up as a neutron star? In that case, everything keeps collapsing under its own extreme weight into an inconceivably compact object. When 3 to 5 solar masses of matter are squeezed into a tiny volume smaller than a pinhead, according to Einstein, space-time in its vicinity is strongly curved. Not even light can escape its immediate proximity. This supercompact object with its infinitely bent space-time environment is nothing else but a black hole. It can be detected only indirectly by measuring the radiation released as matter is falling in.

4.5 Supernovae and Supernova Remnants

Many stars do not live their lives in solitude but spend it together with a "sibling star" in a binary-star system. Within such a system, exchanges of matter between the two stars can repeatedly occur. Gas spirals in from the more massive companion to the lower mass star. Often, one of the two low- or intermediate-mass stars has already become a white dwarf. As a result, the formerly lower mass companion has become the more massive object in the system and can then transfer matter onto the white dwarf. Figure 4.5 illustrates these processes.

This way, a white dwarf gains mass from its companion. If it exceeds the so-called Chandrasekhar limit of 1.4 solar masses, the pressure of

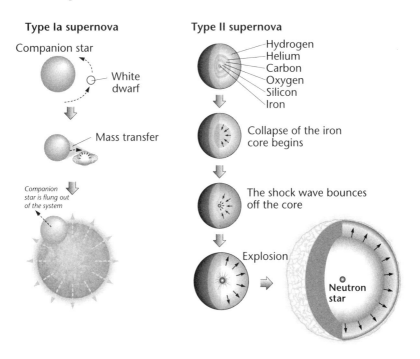

Figure 4.5. The two different mechanisms leading to the explosion of a star or white dwarf. *Left*: Type Ia supernova caused by the explosion of a white dwarf following a mass transfer event. *Right*: A Type II supernova develops at the end of the life of a massive star whose core collapses, thereby triggering the explosion. (*Source*: Peter Palm)

the degenerate electrons that holds the white dwarf together cannot withstand the gravitational force anymore. The white dwarf begins to contract. The temperature stays constant despite the increasing density at the center since the pressure of degenerate gas is independent of temperature. The high density finally leads to renewed powerful nuclear fusion: carbon burning. This nuclear fusion generates energy and heats up the star. Inside a white dwarf, temperature and pressure are actually independent of each other. Hence, the pressure does not continue to increase and the stellar interior is not cooled down by any expansion. Instead, it continues to heat up, in turn causing the nuclear reactions to speed up. This soon leads to an uncontrollable chain of nuclear reactions in which elements up to iron and nickel are rapidly synthesized.

This kind of runaway fusion occurs so fast that the white dwarf has no chance of survival. It explodes in what is referred to as a Type Ia supernova. In this gigantic event, the white dwarf is completely disrupted and the newly synthesized elements, such as oxygen and iron, are dispersed into the interstellar gas. Owing to the lower masses of the progenitor stars, exploding white dwarfs are everywhere to be found in large numbers, making them the main producers of iron in the Universe.

A similar fate awaits a binary-star system in which two white dwarfs collide with each other. For a short while, they become a single object that is twice as massive. Then, too, the Chandrasekhar limit is surpassed, ultimately causing this double object to explode.

The end of the life of a massive star with 8 or more solar masses is likewise among the most spectacular events in the cosmos. The stages of this explosion are depicted schematically in Figure 4.5. At the end of the many nuclear burning stages, a massive star contains a core of iron from which no more fusion energy can be obtained.

At this point, the center of the star is already extremely hot from the previous burning stages. Silicon burning, the last burning stage, occurred at over a million degrees in the center. This is hot enough for so-called photodisintegration to begin. In this process, the iron atoms produced in the core are bombarded with photons and are instantaneously broken apart into protons and neutrons. Billions of years' worth of nucleosynthesis is suddenly undone. Energy is generated when elements fuse together, but energy is needed for the photodisintegration—this energy loss thus leads to a sudden loss of pressure making the core collapse.

During the last burning stages, the star has a stable core held together by the pressure of degenerate electrons. The protons newly released by photodisintegration can then be captured by the many degenerate electrons and transmuted into neutrons and neutrinos. In that way, a neutron-neutrino pair is formed from each proton-electron pair. This signifies an enormous production of neutrinos. The related loss of energy is of central importance to the subsequent explosion of the star.

Before we consider the role of neutrinos further, the processes in the core should first be explained. The absent electrons cause a loss of pressure at the center that lets the core collapse even further. This collapse takes place at immense velocity: at about 70,000 km/s at the outer parts of the core. For comparison, under such conditions the Earth would be compressed into a sphere of 50 km radius within just one second.

The processes in the core happen so fast that the outer layers of the star do not notice any of the changes. Meanwhile, the rapid collapse causes the inner portion of the core to quickly reach the immense density of 10^{17} g/cm^3. The density of nuclear matter is reached, which means that the stellar core is now comparable to a giant atomic nucleus. The collapsing inner region has become a neutron star composed of highly compressed degenerate neutrons left from the photodisintegration. Neutron stars are extremely dense. They typically have 1 to 2 solar masses and a diameter of 10 to 20 km. The Sun's diameter, by contrast, is about 1,400,000 km. One cubic centimeter of a neutron star, which corresponds to roughly the volume of a sugar cube weighing 2 grams, weighs over 10 million tons in comparison.

The neutron star at the center of the collapsing star cannot be condensed any further. Instead, any infalling matter bounces off the "hard" core and is flung back outward again. The outward facing shock wave formed this way travels through the outer parts of the still collapsing inner regions of the star. This leads to a further heating of the matter there, and further photodisintegration. The renewed production of neutrons causes a strong neutron flux by which, within a matter of seconds, many more elements are formed. These are heavier than iron and are called neutron-capture elements. Details of the production of these elements and the observations in metal-poor stars are given in chapter 5. The heat of the collapsing core region needs energy, however, which is now taken from the shock wave. As a result, the shock wave slows down and comes almost to a standstill.

In the meantime a huge accumulation of neutrinos has built up behind the shock wave. This wave of neutrinos coming from behind injects new energy into the stalled shockwave. It is this impulse by the neutrinos that finally leads to the actual explosion and disintegration of the star. Only in this way can the shock wave pick up speed again to blast from the core region into the outer stellar envelope and finally through the whole star. This process tears the star apart. The shock wave pushes the outer stellar layers as well as the newly synthesized elements from the different burning stages out of what was once the star. When the shock wave reaches interstellar space, these newly synthesized elements are thus ejected into the surrounding gas. Interestingly, the neutrinos leave the disintegrating star before the photons. The explosion of a star with 20 solar masses thus has a neutrino luminosity 10 million times higher than its subsequent maximum photon luminosity. This kind of supernova, that is, collapse of the iron core of a massive star, formation of a shock wave, and subsequent disintegration of the stellar envelope, is designated a Type II supernova.

Only after the shock wave has expanded to a diameter of 15 billion km is the gas thin enough for the photons contained inside to finally escape. This is the moment when we can observe such an event as a supernova. While exploding, a supernova shines billions of times more brightly than the Sun for a few days. It becomes as bright as the entire galaxy in which the supernova is occurring. For several months these bursts of energy become the brightest phenomena in the observable Universe, making it possible to observe supernovae in very distant galaxies. The ejected layers of gas form a so-called supernova remnant later on that continues to glow and emit radio waves for some ten thousands of years. Plate 4.C shows the 11,000-year-old Vela supernova remnant as an example. Its gas has been mixing with the interstellar gas since the explosion.

Supernova explosions are the most energetic events in the Universe, irrespective of their type. One can even observe the afterglow of the explosion for weeks on end. The glow is caused by the decay of huge quantities of radioactive nickel (^{56}Ni, with a half-life of about six days), which decays via cobalt (^{56}Co, with a half-life of 78 days) into iron (^{56}Fe). The temporal evolution of their brightness and luminosity can be recorded in the form of a light curve. Its shape reflects these two decay processes that occur in the disrupting stellar envelope.

Figure 4.6 shows example light curves for the two types of supernova explosions, that caused by an exploding white dwarf (supernova Type Ia) and that caused by a collapsing stellar core (supernova Type II). A Type Ia supernova is noticeable only through the radioactive decay of its nickel and cobalt. Accordingly, the light curve drops off faster than for a Type II explosion. For this reason, Type Ia supernovae sustain their maximal luminosity only for a few days. As the shock wave spreads following the core collapse, the Type II supernova will stay brighter for longer because luminous material is pushed outward in this

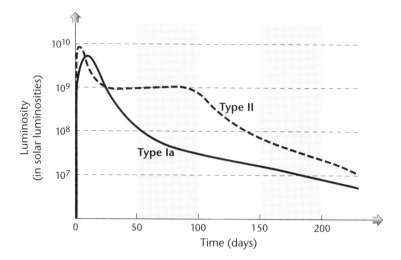

Figure 4.6. Schematic view of the light curves of the two types of supernova explosions. The luminosity of Type Ia supernovae drops rapidly within a few days, while the curve of Type II supernovae fades away slower over the period of a few weeks. (*Source:* Peter Palm)

process. A delayed decrease in luminosity results, as can be seen in Figure 4.6. Core-collapse supernovae consequently shine for many weeks before they fade away.

Depending on the explosion mechanism, the light curve thus has a characteristic shape. This allows observers to identify the type of supernova. The light curve shape is also helpful for constraining details of the explosion. The many difficult and very complex processes occurring during a supernova can be modeled with extremely sophisticated computer simulations. Nonetheless, the theoretical understanding of these explosions is still limited.

Over the past 1,000 years, many supernova explosions in the Milky Way have been recorded. The brightest supernova ever occurred in 1006. It was just 7,000 light-years away and was visible to the naked eye in the sky. Monks at the Abbey of Saint Gall in Switzerland described it in their chronicles. Modern observations of the remnant concluded that this supernova arose from an exploding white dwarf. In 1054, Asian sources mention a very bright supernova whose remnant is the pretty Crab Nebula in the Taurus constellation. Plate 4.C presents this still luminous remnant. Today, we know that it is the result of a core collapse of a massive star that occurred about 6,300 light-years away.

Five hundred years later, the Danish astronomer Tycho Brahe and other astronomers observed another Type Ia supernova in 1572. Just a few decades afterward, in 1604, the famous mathematician and astronomer Johannes Kepler had the great fortune to witness the last supernova in our own Galaxy so far. All these supernovae left behind colorful remnants that are still being observed and studied to the present day. On average, one or two supernova explosions seem to occur in our Milky Way every 100 years.

One more supernova that has to be mentioned in this context is Supernova 1987A. It exploded in 1987 in the Large Magellanic Cloud, the largest satellite galaxy of the Milky Way which is 160,000 light-years away. This event happened only about 30 years ago, and I can still remember it vaguely. In those childhood days, I did not understand anything about exploding stars. But the grown-ups were talking about it, so I gathered that something important had happened. This supernova was, in fact, visible to the naked eye in the Southern Hemisphere and was the closest supernova to us since the Type II explosion in 1604.

Thanks to modern television broadcasting, anyone with a TV was able see the image of that small exotic point of light in another galaxy, whose light had been traveling to us for the previous 160,000 years. This event was an extraordinary opportunity for professional astronomers to examine a supernova "right at their doorstep" in great detail. In particular, the hypothesized immense neutrino flux supposed to drive the shock wave could be experimentally confirmed. Since neutrinos leave the exploding star faster and therefore earlier than the photons, they should arrive on Earth earlier as well. Huge neutrino detectors in Japan and the United States did indeed register those particular neutrinos about three hours before the arrival of the incoming photons that could be observed by the telescopes as light.

When supernova breakouts are discovered in distant galaxies, an additional spectrum is needed in most cases to classify the supernova unambiguously. Spectroscopic observations reveal significant differences in this regard: spectra of Type II supernovae display prominent hydrogen lines, whereas Type Ia objects do not show any hydrogen. A special spectral classification for supernova spectra was accordingly

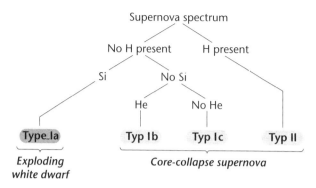

Figure 4.7. Schematic view of the classification of supernova spectra. The existence of hydrogen lines in the spectrum is the main distinction between a supernova of Type Ia and one of Type II. Analyses of core-collapse spectra have shown that two other subgroups exist, Type Ib and Type Ic, which are relatively rare, however. (*Source*: Peter Palm)

introduced. Figure 4.7 depicts this scheme, which also shows additional subgroups of supernova explosions. Their enormous luminosities make both supernova types useful distance indicators as they are still observable in distant galaxies. As a consequence, they are very important for various cosmological studies, such as mapping the expansion history of the Universe.

Indeed, we are indebted to Type Ia supernovae for the astonishing finding that the Universe has been expanding at an accelerating rate for a few billion years by now. For this important finding, the American and Australian astronomers Saul Perlmutter, Brian Schmidt, and Adam Riess were awarded the Nobel Prize in physics in 2011. So-called dark energy is deemed to be behind this accelerated expansion, whose true nature remains unexplained.

4.6 Preliminary Thoughts about Working with Metal-Poor Stars

Our work with metal-poor stars uses the Hertzsprung-Russell diagram to gather information about the evolutionary stage of each new star that is analyzed. Figure 4.8 displays the positions of 130 of the most metal-poor stars in a theoretical Hertzsprung-Russell diagram designed for this purpose. This version of the diagram plots surface gravity against temperature. Both these measurable quantities can also be computed from simulations of stellar evolution. Such computed evolutionary tracks have been added to the figure for comparison.

These theoretical tracks, called isochrones (from Greek: *iso* = equal, *chronos* = time), refer to the evolutionary paths of same-aged stars of different masses. Different metallicities can also be selected when computing these artificial stellar populations. In this case, 12-billion-year-old stars with metallicities from one-tenth to one-thousandth of solar iron abundance have been modeled. The main sequence is the set of almost horizontal lines toward the bottom. Depending on metallicity, the turn-off point occurs between 6,000 and 6,700 K. The red giant branch leads from there diagonally upward to the right. The effect of metallicity on stellar temperature can be clearly read off from the various curves. The

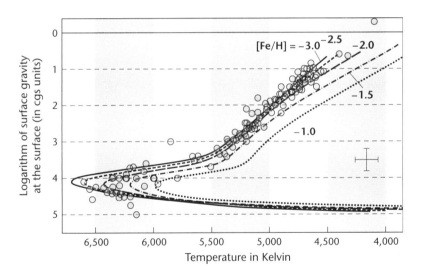

Figure 4.8. Hertzsprung-Russell diagram depicting ~130 of the most metal-poor stars known along with five theoretical curves of stellar evolution, based on different metallicities. The red giants have surface temperatures ranging from 4,000 to 5,500 K, whereas the main-sequence stars are somewhat hotter at 5,800 to 6,600 K. All the stars follow the metal-poor evolutionary track located further to the left, even though some data points do deviate. So-called error bars that indicate the minimum error in the measurement are shown to the right of the tracks. Within those errors, all the data points agree well with the theoretical tracks. (*Source*: Peter Palm; data from Frebel, *Astronomische Nachrichten* 331 (2010): 474–488)

more well-separated metal-poor stars are hotter and therefore bluer than more metal-rich ones. Hence, the turn-off point of the most metal-poor isochrone is considerably shifted leftward to hotter (surface) temperatures. On the giant branch, any differences with decreasing metallicity become less and less pronounced, however.

As can also be seen in the figure, all metal-poor stars are assembled either near the turn-off point or on the red giant branch. How can we use this information to find out more about the masses of these stars? The Sun has a life expectancy of "just" 10 billion years, which means, after 10 billion years, it would not be on the main sequence anymore but in the region of the white dwarfs. Since all metal-poor stars are located near the turn-off point and on the red giant branch after an estimated 12 billion years, their masses have to be lower than that of the Sun. It can thus be estimated that the near turn-off stars probably have about 0.6 and the red giants about 0.8 solar masses.

Figure 4.8 shows how diverse the known metal-poor stars are: main-sequence stars, ones that are already past the turn-off, and red giants. Considering that the main sequence is by far the longest stage in stellar evolution, one would have expected that most known metal-poor stars would be main-sequence stars. Yet, a different picture emerges. In detailed chemical abundance studies, more red giants have been examined than main-sequence stars. How can this be?

This curiosity reflects an important selection effect. Red giants shine brighter than main-sequence stars, as they are located further up in the Hertzsprung-Russell diagram. This behavior is independent of whether an observer or theorist plots a Hertzsprung-Russell diagram. Compared to a main-sequence star, the intrinsically higher luminosity of a giant means that it is still observable even when far away. Having access to a larger volume this way, the greater the chance that particularly metal-poor stars will be found. These are precisely what stellar archaeology is looking for and why substantially more metal-poor red giants are observed and subsequently analyzed. This kind of analysis is described in chapter 7.

Red giants have another advantage. When searching for metal-poor stars, stellar surface temperature has to be taken into account. Having the same metallicity, the absorption lines to be measured in the stellar spectrum of cooler red giants are more prominent than in the spectrum of a hotter main-sequence star. This is especially important for most metal-poor stars. Their lines are weak and quickly fade into the noise, becoming unrecognizable as such. With red giants this happens at substantially lower metallicities, so it often turns out to be easier to measure features in the spectra of those objects. In the end, it is more promising and efficient to choose red giants from among the list of candidates for the next observations and to leave any main-sequence stars aside.

The positions of metal-poor stars in the Hertzsprung-Russell diagram indicate the evolutionary stage of each star. This knowledge is important when assessing whether the composition of the stellar surface has been changed by the star itself. The aim of stellar archaeology is to examine the chemical conditions shortly after the Big Bang. Metal-poor stars serve as long-term carriers of this early material because the composition of their outer atmosphere still resembles the composition of the gas cloud at the time of their birth in the early Universe. One direct consequence of this central assumption is any process that alters the chemical composition of surface makes the star useless to stellar archaeology. In those cases, it is impossible to draw correct conclusions about those early chemical conditions.

For example, the stars at the upper end of the red giant branch have already started to dredge up matter from their interiors to the surfaces through mixing processes. Abundance analyses of such stars reveal that these changes are, in fact, significant and thus important to our understanding of stellar evolution and the associated nucleosynthesis processes. But it is exactly these stars we do want to avoid in stellar archaeology. Consequently, it is an important task to establish how far up on the red giant branch a star is sitting to avoid this issue. A star's position in the Hertzsprung-Russell diagram, such as in Figure 4.8, is helpful in this respect. Unlike even cooler giants (not shown in the figure), all the studied red giants are hardly affected by this "self-enrichment" of the surface. For this reason, stars that are not too advanced in their evolution, that is, main-sequence stars and not too cool red giants, are best for stellar archaeology.

Besides this internal mechanism of surface alteration, other external possibilities exist that can change the composition of the stellar surface layer. Main-sequence stars and giants alike are affected but the Hertzsprung-Russell diagram is of no help anymore. It is certainly possible for a star to collect tiny amounts of interstellar gas over the course of its long life, thus chemically contaminating its atmosphere. Luckily, metal-poor stars in the halo of the Milky Way have relatively large velocities compared to the gas, making it difficult for them to accumulate matter this way. Traveling at high speed in a car, one can hardly pick up something on the side of the road, whereas it would be easy to grab something at walking pace. Hence, these two factors do not pose a major problem for stellar archaeology.

Finally, a certain amount of sedimentation does take place inside any star throughout the billions of years of its lifetime. Particularly during the main-sequence stage, atoms on the surface can sink down—very slowly, of course—in the direction of the center. This process can also change the chemical composition of the surface because elements of different weights sink at different rates. Fortunately, these effects are approximately the same for all stars and can theoretically be roughly quantified as well. Yet another process leads to an alteration of the composition of the outer surface, too. Mass transfer between stars in a binary system is a curious exception in this regard, though. As will be explained in chapter 5, this process, however, does not render the altered recipient star entirely useless to stellar archaeology.

NEUTRON-CAPTURE PROCESSES AND THE HEAVIEST ELEMENTS

Nucleosynthesis of the chemical elements up to iron is every star's source of energy. No more energy can be gained from fusing iron, so heavier elements cannot be produced this way. But how could the many heavy elements from the bottom part of the periodic table have formed? Their creation must have occurred through processes other than nuclear fusion. Paradoxically, multiple processes involving radioactive decay are what ultimately allow these heavier elements to be generated. The synthesis of elements heavier than iron takes place stepwise in a repeat procedure of so-called neutron-captures that builds up heavy nuclei.

In a neutron-capture process, an already available "seed nucleus," for example a carbon or iron nucleus, is bombarded with neutrons. In the case of iron, an extremely neutron-rich iron isotope initially forms. Since the proton number has not changed, the nucleus remains an iron nucleus but is unstable due to the added neutrons. This simply means that the nucleus wants to get rid of its excess neutrons again. Decay processes make this happen, whereby new stable isotopes of different elements form. When a neutron spontaneously transmutes into a proton (β-decay), a neutron-rich iron nucleus, for instance, transforms into an element of higher atomic number, namely, cobalt. This process is depicted in Figure 5.1.

Iron nuclei or the nuclei of similar elements need to be bombarded with countless neutrons in order to generate one heavier element after the next. This process works all the way to the production of very heavy elements, such as bismuth, which has 83 protons and 126 neutrons. Technically, bismuth is not a stable element because it decays within a period of 19 trillion years (that corresponds to about 1 million times the

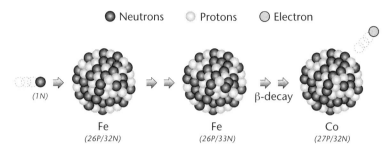

Figure 5.1. An iron atom (a "seed nucleus") transmutes into a heavier cobalt atom after neutron bombardment and subsequent β-decay. (*Source*: Peter Palm)

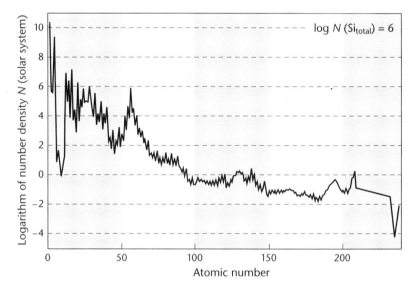

Figure 5.2. The Sun's element-abundance pattern (normalized here to silicon, not hydrogen). It describes the state of chemical evolution 4.6 billion years ago. (*Source*: Peter Palm; data from Lodders et al., *Astrophysical Journal* 591 (2003): 1220–1247)

age of the Universe) into thallium by α-decay. The heaviest truly stable element is lead, with 82 protons and 125 to 127 neutrons. Figure 5.2 shows the abundances from hydrogen to lead in the Sun, as one example of the complete cosmic element production.

Since the timescales over which these neutron-captures onto seed nuclei occur play a fundamental role in the formation of the heavy elements, a distinction is made between the slow "s-process" and the rapid "r-process." These two processes take place in very different astrophysical sites where the necessary neutron fluxes can be generated in different ways. They will be examined more closely in the following.

5.1 How Neon Lamps Relate to Giant Stars—Element Synthesis in the S-Process

Toward the end of its life, a star of less than 8 solar masses ejects all its gaseous layers. They are later observable as a beautiful planetary nebula. Chapter 4 described these processes at length. Detailed chemical analyses of such nebulae have shown that they contain many heavy and exotic elements. For example, neon can be found there, as can germanium, selenium, bromine, krypton, xenon, and rubidium. We usually associate most of these elements with neon signs and fluorescent tubes. As a gas in a lamp, neon glows orange-red, krypton white, and xenon bluish purple. We are indebted to giant stars for these colorful effects. Those elements were produced a long time ago by an s-process in the outer stellar layers prior to ejection of the external envelope. It was long before the elements were incorporated into our own planet when the Solar System and Earth were forming, and long before they were used in neon advertising.

The s-process takes place in intermediate-neon stars of roughly 2 to 8 solar masses that are in their final evolutionary stage on the asymptotic giant branch. This final stage lasts less than 1% of the star's lifetime. Having turned into a cool asymptotic giant branch star, it possesses a huge outer atmosphere in which the transfer of energy to the surface takes places only by convection. This envelope, broadened to some 100 solar radii, pulsates at regular intervals causing it to dredge up any fresh s-process elements upward from deeper layers. The star resembles a giant concrete mixer. This mixing of course alters the chemical composition of the star's surface. Strong stellar winds then sweep these new elements off the surface and release them into the interstellar medium. Intermediate-mass stars can lose up to one solar mass of matter every 100,000 years this way. Consequently, the asymptotic giant branch stars are extremely important participants in the chemical enrichment of a galaxy and therefore also of the Universe.

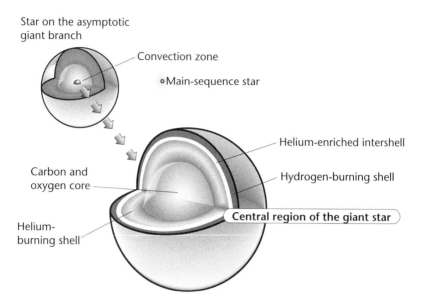

Star on the asymptotic giant branch

Convection zone

Main-sequence star

Helium-enriched intershell

Hydrogen-burning shell

Carbon and oxygen core

Central region of the giant star

Helium-burning shell

Figure 5.3. The s-process takes place above the helium-burning shell in an intershell enriched in helium within an asymptotic giant branch star. This pulsating region is beneath the hydrogen-burning shell at the lower boundary of the convective zone. The core consists of carbon and oxygen. Assuming this core to be just 1 cm in size, the surface would be 500 m away. (*Source*: Peter Palm)

Located underneath the convective stellar envelope is the hydrogen-burning shell, and below that the helium-burning shell. As Figure 5.3 illustrates, the s-process takes place in a regularly pulsating intershell between the two burning shells. Many of the nucleosynthetic details of the s-process are well understood theoretically, although some details still remain unclear. An example is the need for a detailed model of the neutron source in the intershell.

Basically, a neutron source arises if carbon isotopes (^{13}C) or neon isotopes (^{22}Ne) capture α-particles, that is, helium nuclei. A neutron is released at each capture. It generates a relatively low but long-lasting neutron density of at least $\sim 10^8$ neutrons/cm^2/sec, depending on the stellar mass. Nevertheless, enough neutrons are known to be provided by these reaction processes. The neutrons are available particularly at the bottom of the convection zone, where they are repeatedly incorporated into the available seed nuclei, such as iron, over the course of millennia.

After a new neutron has been captured, the newly formed radioactive, hence, unstable, atom decays again. This happens before it is bombarded by another neutron. The steady capturing of neutrons and subsequent decaying leads, slowly but surely, to the buildup of a heavier stable nucleus. About half of all isotopes heavier than those of iron are synthesized in the s-process. Figure 5.4 illustrates this buildup schematically. However, many of these isotopes are produced not only by the s-process but also by the r-process. But a few isotopes are exclusively made in the

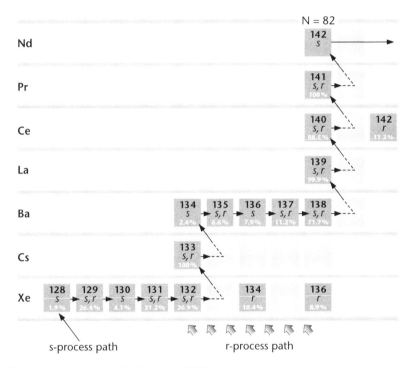

Figure 5.4. Example synthesis paths of different neutron-capture elements by the s-process and r-process. The darker boxes are stable elements, lighter boxes are unstable ones. The upper number indicates the atomic mass of the isotope. The percentages indicate the fraction of that isotope of the element. It is furthermore shown whether the isotope can be generated solely by the s-process or solely by the r-process, or by both. For example, in the s-process, a xenon atom is bombarded with one neutron after another. When an unstable isotope forms, it β-decays into a cesium isotope. After another neutron-capture it decays into barium. Stable barium isotopes, in turn, capture other neutrons before the heaviest unstable one decays into a lanthanum isotope. The capturing and decaying continues up to neodymium, and even further, which is not depicted. Some xenon and cerium isotopes can be produced only by the r-process from the decay of substantially heavier neutron-enriched isotopes, which are not drawn for space reasons. (*Source*: Peter Palm; data from Sneden et al., *Annual Review of Astronomy and Astrophysics* (2008), pp. 241–288)

s-process, such as ^{86}Sr (strontium), ^{96}Mo (molybdenum), ^{104}Pd (palladium), and ^{116}Sn (tin).

In absolute quantities, the neutron-capture elements synthesized in the s- and r-processes are extremely small. Typical neutron-capture elements, such as strontium, barium, and europium, are produced in quantities about 1 million times less than elements in the iron group. These elements are enormously important all the same, not just for studying the beginnings of chemical evolution in the cosmos but also regarding the origins of us human beings. Selenium, an essential trace element for the thyroid gland, is an s-process element made in asymptotic giant branch stars.

The total abundance of new elements depends not only on the neutron flux but also on the number of available seed nuclei inside a star. Compared with the Sun, metal-poor stars have few iron atoms because they formed from gas of low metallicity. Consequently, there are many more neutrons available per seed nucleus in a metal-poor star. While the neutron source is operating, the fewer available seed nuclei catch enough neutrons to go through the entire s-process chain. Elements as heavy as bismuth and lead can thus be produced in large quantities. In contrast, inside more metal-rich stars there are fewer neutrons available for every seed nucleus. When the neutron source dwindles, the neutrons have to be shared among too many seeds and the s-process usually grinds to a halt before having produced larger quantities of the heaviest elements.

But how do the heavy elements we observe get inside our metal-poor stars? Evidently, stellar winds of such far-advanced asymptotic giant branch stars enrich the gas with s-process elements before the formation of any new metal-poor stars. However, owing to a major shortage of iron, the s-process probably never took place at the earliest times—that is, during the era of the most metal-deficient stars. At least trace amounts of nuclei such as iron are required within the stellar atmosphere itself to act as seed nuclei. It is thus unclear when the s-process occurred in the Universe for the very first time. Moreover, it could have taken up to a billion years before the significant numbers of asymptotic giant branch stars began synthesizing s-process elements.

The chemical evolution began only slowly in the early Universe, step by step, with the production of elements up to iron in short-lived massive stars. Sometime later, s-process elements made in intermediate-mass stars of less than 8 solar masses were added. The longer lifetimes of these stars led to a delay in the enrichment of the interstellar medium by their stellar winds. This "tardiness" is, in fact, reflected in the abundances of metal-poor stars. Only from a specific higher, hence "delayed," metallicity onward does one find metal-poor stars with abundance patterns of neutron-capture elements associated with the s-process.

A particular class of metal-poor stars can actually provide more information about the beginnings of s-process nucleosynthesis. Some otherwise "quite normal" metal-poor stars possess unusually large amounts of carbon and s-process elements: more than 10 times compared to iron. Where do these immense abundances of s-process elements come from?

As extensive observations have demonstrated, these so-called s-process stars are members of binary-star systems. In such partnerships two stars closely orbit each other. This can be established by measurements of stellar radial velocities, that is, the stellar velocities in the direction of the observer's line of sight. Usually only the brighter object can be observed. If the velocity varies over the course of months to years, then the star is moving around a by now unseen companion star. The masses of these two chemically identical stars born from the same gas cloud differ, though. The more massive companion already became an inflated giant a long time ago. At some point, given its shorter life span, it had moved onto the asymptotic giant branch and produced s-process elements in its inner region that were subsequently dredged onto its surface. Since carbon played an important role in the s-process, that is, as neutron source, large amounts of carbon reached the surface together with the s-process material.

In a cosmic balancing act, the giant star then donates a portion of its outer atmosphere to its lower mass companion star. This "gift" is the reason why s-process elements can be found in the spectrum of an otherwise rather inconspicuous metal-poor star. By now, only the light of the low-mass star is observable. The erstwhile giant has long since evolved into a faint white dwarf. Unlike before, it does not outshine its companion anymore.

This special enrichment procedure explains observations of unusually large amounts of s-process elements and carbon in some metal-poor stars. The s-process elements were produced not by the observed metal-poor star itself but by its fellow binary-star companion, which had manufactured it while in the asymptotic giant branch stage.

Not surprisingly, observations of the most metal-poor s-process stars have shown that they possess immense abundances of lead, just as predicted. Most of the lead used in diving belts or radiation-proof aprons, for example, has been produced by the s-process.

For astronomers and nuclear physicists, it is very fortunate that s-process stars exist. These metal-poor stars act as carriers of s-process material, making it possible to study the s-process and its element production very closely. Whereas only a small number of s-process elements are detectable in planetary nebulae, up to 20 are found in metal-poor s-process stars. Detailed abundance analyses have contributed significantly toward improving theoretical models of nucleosynthesis as well as our understanding of the astrophysical production site inside evolved asymptotic giant branch stars.

5.2 Thorium, Uranium, and R-Process Element Synthesis

Whereas the s-process operates in environments with relatively low neutron densities inside asymptotic giant branch stars, the r-process requires a substantially stronger neutron flux. The required extreme conditions can be achieved only in a supernova, or during the merger of two colliding neutron stars. A promising candidate for the astrophysical site of the r-process are massive stars of 8 to 10 solar masses or those with more than 20 solar masses that explode as Type II supernovae. Since the first phase of chemical evolution was initially driven by supernova explosions, it would follow that the r-process was already operational during the earliest epochs of the Universe. This is supported by the existence of metal-poor stars exhibiting r-process elements. But it has not yet been possible to determine the specific production site because many details of r-process nucleosynthesis are still insufficiently understood. Nonetheless, the rough characteristics of the r-process have been known for many decades. For example, the r-process operates on timescales of just two to three seconds! During that short time, seed nuclei, such as carbon or iron nuclei, are heftily bombarded with neutrons.

Supposedly, ~10^{22} neutrons/cm^2/sec are required. This is a huge number and two seconds is an astonishingly short time—taking a deep breath takes longer.

The main reason for the speed of the r-process is that whenever a nucleus acquires an extra neutron it wants to immediately decay. This procedure occurs in the s-process. But if other neutrons are promptly added before the nucleus has decayed, then a very neutron-rich isotope can temporarily be produced. Those large, unstable nuclei decay, which leads to the production of half of the isotopes of all the heavier elements in the periodic table (with the s-process being responsible for the other half).

The r-process is thus like trying to walk up a downward-moving escalator. The disappearing escalator steps resemble nuclei decaying one after another. In order to build a heavy nucleus, you would have to run up the escalator pretty fast, faster than it is moving down, otherwise you would not get to the top. If you are not fast enough, you will just be jogging in one spot. This would then correspond to the s-process.

For various nuclear physics reasons, the r-process cannot produce arbitrarily heavy elements. When a nucleus gets too heavy during the neutron bombardment, it fissions instantaneously into lighter nuclei. In general, after the neutron flux ceases, the heaviest possible isotopes have about 100 protons and up to 160 neutrons. They are radioactive and decay along an extensive decay chain (mostly by α-decays) into many different isotopes, finally ending up as lead. Lead has 82 protons and about 100 neutrons, depending on the isotope. Built this way are not only the heaviest stable elements but also the heaviest long-lived radioactive nuclei, thorium (^{232}Th), with a half-life of 14 billion years, and uranium (^{238}U), with a half-life of 4.7 billion years. These long half-lives are useful for measurements of cosmic timescales.

All in all, the r-process leads to very characteristic ratios between the abundances of the heavy elements. This r-process signature is readily identifiable. The s-process also has a characteristic pattern, but it differs markedly from that of the r-process. These patterns are described more closely in what follows.

The abundances of chemical elements in the Sun are the cumulative product of 9 billion years of chemical evolution, from the Big Bang until the Sun's formation about 4.6 billion years ago. Accordingly, the heavier

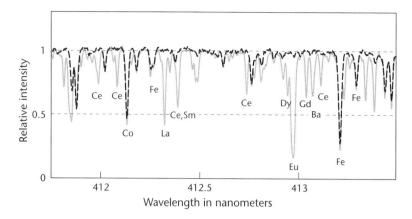

Figure 5.5. Portions of normalized high-resolution spectra of two metal-poor stars. One is a normal metal-poor star (black dashed line) and the other is an r-process star (gray line). The latter shows many absorption lines not found in the normal metal-poor star. They are attributable to the existence of neutron-capture elements in the r-process star. The strong r-process europium line at 412.97 nm is clearly visible. (*Source*: Peter Palm; spectra from Anna Frebel's private collection)

elements in the Sun are a mixture of isotopes manufactured by the s- and r-processes. Since the s-process pattern is very well known theoretically, it can be subtracted from the measured solar abundance pattern. What remains is the r-process component. This solar r-process pattern can then be compared to theoretical predictions of r-process nucleosynthesis.

For a long time the Sun's r-process pattern was the only means by which to gather empirical data on the r-process operating in the cosmos. In 1995, the first extremely metal-poor "r-process star" was found, though. Absorption lines of about 70 elements of the periodic table were measured in the spectrum of CS 22892-052. They include not only the usual lighter elements such as carbon, oxygen, magnesium, sodium, titanium, iron, and nickel, but above all neutron-capture elements, such as strontium, barium, europium, gadolinium, dysprosium, praseodymium, and osmium, to name just a few. By now, more such stars have been discovered, and some of them present even thorium and uranium. Figure 5.5 compares the spectra of an r-process star with an ordinary metal-poor star. More elements than those found in r-process stars can only be detected in the Sun.

Judging from its low iron metallicity, this metal-poor r-process star formed in the early Universe as part of the first few stellar generations. As no s-process nucleosynthesis took place during those early epochs, this star likely emerged from a gas cloud that had already been enriched with r-process elements before it was born. This means that metal-poor r-process stars are carrying the direct chemical fingerprint of an r-process event that occurred in a star of the previous generation. We can thus conclude that r-process events took place already soon after the Big Bang.

Moreover, it is widely established by now that the observed abundances of heavy elements in metal-poor r-process stars agreed *exactly* with those of the solar r-process pattern, up to a simple scaling factor. This is the case particularly for elements heavier than barium (atomic number 56). As Figure 5.6 shows, the scaled patterns of an r-process star and of the Sun are identical within the observational uncertainties.

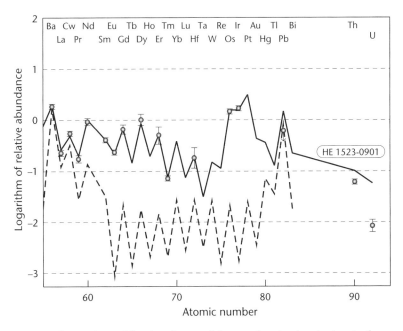

Figure 5.6. Comparison of the abundances of elements heavier than barium in the r-process star HE 1523–0901 (circles) and the scaled solar r-process pattern (line). The agreement is astonishingly good. The solar s-process (dashed) is given for comparison. It does not agree with the observed abundances. (*Source*: Peter Palm; data from Frebel et al., *Astrophysical Journal Letters* 660 (2007): 117–120)

How can ancient stars and the Sun share the same element pattern, though? In view of the fact that the Sun was born some 9 billion years later than ancient metal-poor r-process stars, this is an astonishing discovery. It can only be explained as follows: The r-process is a universal process, at least when it comes to the formation of the heaviest elements above barium. Wherever and whenever it occurs, these heaviest elements are produced in exactly the same proportions. There is just one "secret recipe" that nature repeatedly uses in exactly the same way. It is the task of astronomers and physicists to figure out the ingredients of this recipe to eventually understand the r-process. Since there are severe limits to synthesizing heavy r-process elements in the laboratory, if at all, researchers have been using the r-process stars as their cosmic laboratory to test nuclear physics and astrophysics theories on the origin of these elements.

Chemical analyses of metal-poor r-process stars have thus opened up new means by which to study r-process nucleosynthesis directly and, at the same time, to gather important information about where the r-process might take place in the cosmos. An immediate conclusion would be that a supernova produced the r-process elements and expelled them into the interstellar gas that was later incorporated into the r-process star. Interestingly, the solar r-process element abundances do not actually permit this conclusion because countless generations of stars have enriched the Sun's native gas cloud with s- and r-process elements. Other sites with longer timescales could well have contributed r-process elements. This is where neutron star mergers have to be considered, despite the fact that the time for the two neutron stars to coalesce is much longer than a massive star lifetime, and thus seemingly at odds with the early existence of metal-poor r-process stars. Regardless, these examples illustrate how metal-poor r-process stars assist us in a unique way in constraining the details of the origin of the heaviest elements.

Lighter elements up to iron and nickel exhibit a well-defined and continuous trend in their cosmic evolution, as can be seen in Figure 5.7. This is attributable to a thoroughly mixed, homogeneous star forming gas going back to very early times. In contrast, the evolution of neutron-capture elements paints a different picture. It is anything but well-defined and exhibits a vast amount of scatter. Very different amounts of r-process elements (and neutron-capture elements in general) must have been synthesized in each supernova, particularly in the early Universe.

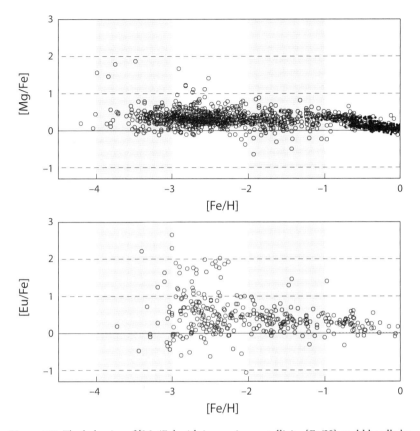

Figure 5.7. The behavior of [Mg/Fe] with increasing metallicity [Fe/H] could hardly be more different from that of [Eu/Fe]. Magnesium exhibits a very well-defined trend (although there are exceptions), which can be ascribed to its synthesis always occurring the same manner. By contrast, a clear trend for europium is discernable only at substantially higher metallicities because the gas had already been very well mixed during later epochs. Therefore, europium was likely produced in rather variable quantities in the early Universe. The definition and description of logarithmic abundances [Fe/H], [Mg/Fe], and [Eu/Fe] are given in section 7.3 (p. 166). (*Source*: Peter Palm; data from Frebel, *Astronomische Nachrichten* 331 (2010): 474–488)

But this is probably not the sole explanation for the somewhat convoluted abundances of the neutron-capture elements. There is much speculation going on about whether other processes exist, for example, that exclusively produce lighter neutron-capture elements and none or only little of the heavier ones. The topic remains a subject on ongoing research, both observationally and theoretically. In conclusion, it is clear from the different behaviors of light and heavy elements that the production of all neutron-capture elements is entirely independent of the synthesis of the elements up to iron.

5.3 Cosmo-Chronometry: The Oldest Stars

Although globular clusters are not nearly as deficient in metals as the most metal-poor stars in the halo of our Milky Way, they are still among the oldest objects in the cosmos. The age of a globular cluster can readily be determined since all the stars in a cluster formed at the same time from the same material but differ from each other only in mass. The corresponding dating method, which is based on the Hertzsprung-Russell diagram, is described in chapter 6. As the stars in the Universe cannot be older than the Universe itself, the age of globular clusters sets a natural lower limit for the age of the Universe.

For a high school project, I once obtained ages of various globular clusters with this method. I was able to convince myself that the cluster ages are indeed around 12 to 14 billion years. At that time, the generally accepted age of the Universe was still 15 billion years.

In 2003, the Universe's age was determined with great precision for the first time and found to be 13.7 billion years old. This age is derived from an analysis of the cosmic background radiation left behind by the Big Bang, as measured by NASA's WMAP satellite. New results from the Planck space satellite flown by the European Space Agency have recently slightly revised the Universe's age to 13.8 billion years. This age does not necessarily pose a problem for globular clusters suddenly appearing to be older than the Universe. Any age measurements of star clusters have often uncertainties of a few billion years. New redeterminations of globular clusters' ages yield values of 10 to 12 billion years by now, mostly due to increased data quality. These values agree well with the Planck age of the Universe. Nonetheless, clusters remain among the most ancient objects known.

Individual most metal-poor stars in the halo of our Milky Way formed at a time when the Universe had experienced very little metal pollution from supernova explosions, that is, very soon after the Big Bang. For this reason, these stars, just like globular clusters, are assumed to be almost as old as the Universe. Unfortunately, the ages of individual stars

tion lines in the optical spectrum that become measurable in r-process stars.

In 2001, a second r-process star, CS 31082-001, likewise an extremely metal-poor halo star, was discovered in the same survey. Not only thorium lines but also the one extremely weak uranium line available in the entire optical spectrum between 300 and 800 nm was detected in its spectrum. For the first time, a uranium abundance could be measured in a metal-poor star leading to an age of 14 billion years based on the uranium : thorium chronometer.

Interestingly, in the case of CS 31082-001, only the uranium : thorium chronometer yielded a physically meaningful age. Other abundance ratios, such as thorium : europium, provided no useful age information, because a *negative* age was found. As even astronomers cannot look into the future—a negative age, if correct, would imply it. This is a classic example of an "unphysical" finding. It thus immediately became clear that many details in the r-process and its astrophysical production site are still in dire need of further research. For that, other r-process stars are required, particularly the ones for which more than one chronometer abundance ratio (i.e., thorium : europium) can be obtained. Only this way can the details of r-process nucleosynthesis as well as its puzzling exceptions be understood one day.

These two r-process stars were serendipitously discovered in their given samples, together with several somewhat more metal-rich and substantially less r-process-enhanced stars. In an effort to systematically search for more of these rare stars, a large observing campaign was started in 2001 using one of the four 8-m Very Large Telescopes of the European Southern Observatory (ESO). The aim was to collect spectra of 350 stars from the Hamburg/ESO Survey to examine them for the europium line at 412.9 nm. A strong europium line in the spectrum is the best indicator, the "smoking gun," that an r-process star has been found because all isotopes of this element are almost exclusively synthesized by the r-process.

After a few years, a dozen r-process stars with very large europium abundances of over 10 times more europium than iron were indeed identified by this effort, along with other stars with somewhat lower Eu abundances. All in all, this study showed that 5% of metal-poor stars are extreme r-process stars. Most of these new discoveries could be dated by means of the thorium : europium chronometer to ages similar to those of the first two stars. However, uranium could not be detected in any object. It was impossible to obtain the extremely high data quality required for a uranium measurements of these rather faint stars.

A uranium detection is possible only if an r-process star meets a number of requirements:

1. First of all, the star needs to contain a large amount of r-process elements compared to iron, at least 50 times or more.
2. The star should be as bright as possible to be able to obtain the extremely high-quality data needed within a reasonable amount of telescope time. A stellar V magnitude of 12 or less is preferred.
3. The star has to be extremely metal-poor so that the absorption lines of other lighter elements, such as iron or titanium, do not unnecessarily obscure the lines of the r-process elements in the spectrum.
4. The star should be very cool and situated at the middle or upper end of the red giant branch. Surface temperatures between 4,500 and 5,000 K are desirable because then the absorption lines are more prominent than those in hotter stars. Only then does the tiny uranium line emerge slightly stronger and possibly becomes detectable.
5. Finally, the star should have a low carbon abundance. Directly next to the uranium line lies a relatively strong molecular carbon line that, when the carbon abundance is large, completely blends and obscures the uranium line. A low abundance of carbon is also advantageous for measuring the thorium abundance. Several molecular carbon lines are lurking in the spectral region of the strongest thorium line as well. However, since the thorium line is stronger than the uranium line, the risk of severe blending is substantially less.

In light of all these requirements, candidates for age determinations by means of a uranium chronometer are extremely rare. The first r-process star, CS 22892-052, would have been the perfect candidate, but because it is also carbon-rich, no uranium measurement is possible. Hence, there are still only three metal-deficient stars in which uranium has been detected, although one of them has a much less secure measurement. In contrast, the star HE 1523-0901 is almost a textbook case.

In 2005, this bright halo star was discovered in the Hamburg/ESO Survey. It was one of the sample objects forming the basis of my PhD thesis. Soon after, I was able to measure the largest abundances of r-process elements known—thorium and uranium among them. The spectral region with the tiny uranium line in HE 1523-0901 is shown in Figure 5.8, in comparison to CS 31082-001. HE 1523-0901 is the first star for which more than one chronometer has been applied. With seven different chronometers, its age was established at 13.2 billion years. This averaged age agrees well with the age of the Universe of 13.8 billion years as calculated from the Planck satellite data.

Individual stellar age determinations are often subject to large measurement uncertainties, stemming from not only the element abundance determination procedure applied to the spectrum but also the calculated initial values of the element production in the r-process. They can be a few billion years, depending on which chronometer is used.

Despite these uncertainties in individual age measurements, the ever important question of the age of the oldest stars can finally be answered with the entire group of extremely metal-poor r-process stars and all their chronometers: they are almost as old as the Universe itself. Stellar ages of 12 to 14 billion years have been measured for all of these stars. This fact somewhat compensates for the uncertainties of individual stellar age determinations.

What follows is one of the most crucial points to address with stellar archaeology: are all metal-poor stars in fact almost as old as the Universe? This is actually one of the assumptions upon which stellar archaeology rests. The answer to this question is generally "yes." The reason is that the abundances of lighter elements in r-process stars do not differ in any way from those in ordinary metal-poor stars. R-process stars are ordinary metal-poor stars that merely received a supplementary r-process elements portion during their formation. As long as stars have similarly low metallicities, it is thus likely that they are all of similar age. This conclusion also justifies the assumption that metal-poor stars would have masses of 0.6 to 0.8 solar masses. Only stars with such low masses can survive for billions of years from soon after the Big Bang all the way to the present day.

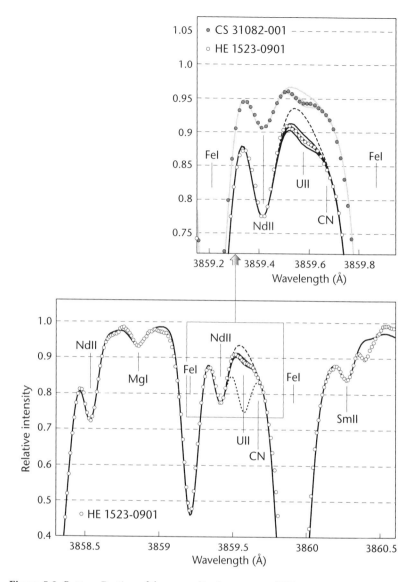

Figure 5.8. *Bottom*: Portion of the normalized spectrum of HE 1523-0901 around the uranium line at 385.9 nm. The open circles represent the observed spectrum. The solid line indicates the best fitting synthetic spectrum. The dotted line indicates how strong the uranium line would have been if uranium had not been decaying over the past 13 billion years. *Top*: Detail view on the uranium line. HE 1523-0901 is compared with CS 31082-001, the other "uranium star." CS 31082-001 is very similar to HE 1523-0901, but 200 K hotter. Accordingly, its absorption lines are less prominent than those in HE 1523-0901, despite the nearly identical abundances. The solid lines represent three synthetic spectra with different uranium abundances. The middle one fits best and thus yields the uranium abundance. The dashed line shows what the spectrum would look like if no uranium were present in the star. This synthetic spectrum is not in agreement with observation. (*Source*: Peter Palm; reproduction of the spectrum from Frebel et al., *Astrophysical Journal Letters* 660 (2007): 117–120)

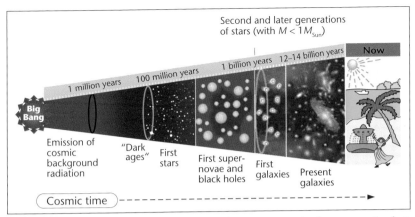

Plate 1.A. The evolution of the Universe: from the Big Bang to the first stars to the first galaxies and finally to life on Earth. (*Source*: Peter Palm)

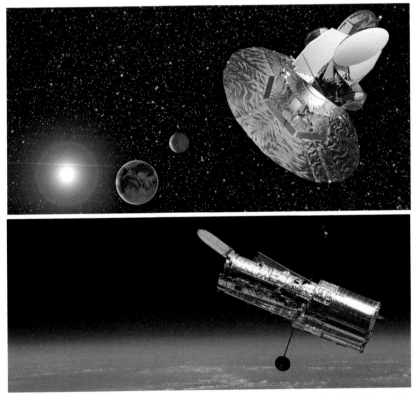

Plate 1.B. *Top*: Artist's impression of the Wilkinson Anisotropy Probe (WMAP) satellite in outer space. The Sun, Earth, and Moon are shown in the background. *Bottom*: The Hubble Space Telescope in orbit around Earth. It was photographed from the Discovery Space Shuttle. (*Sources*: *Top*, WMAP #990387, NASA/WMAP Science Team. *Bottom*, STS-82 Crew, STScI, NASA)

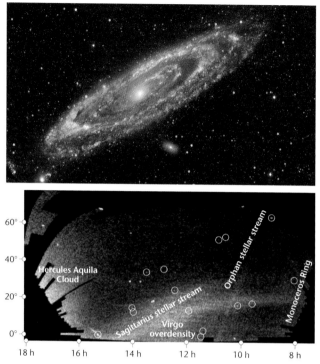

Plate 1.C. Andromeda—the sister galaxy of our Milky Way. A larger dwarf satellite galaxy is visible in the foreground. (*Source*: NASA/JPL-Caltech)

Plate 1.D. The so-called Field of Streams showing the stellar density in the halo of the Milky Way. The bottom part around 0 degrees points into the plane of the Galactic disk, whereas the upper part at higher latitude points outward into the halo. The brighter regions signify high concentrations of stars resulting from extensive streams of torn apart dwarf galaxies. The names and locations of the streams are indicated, as are several faint dwarf galaxies that orbit the Milky Way as satellites. The black areas along the left edge depict a lack of data in those regions. (*Source*: Peter Palm; reproduction of an illustration in Belokurov et al., *Astrophysical Journal Letters* 642 (2006): 137–140. With the kind permission of Vasiliy Belokurov)

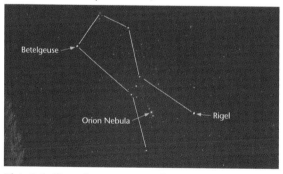

Plate 3.A. The red supergiant Betelgeuse in the Orion constellation. Betelgeuse is the second brightest star in Orion after the white-blue shimmering Rigel. In the Northern hemisphere, Orion can be seen in the wintertime. (*Source*: Peter Palm; reproduction of a photograph by Wolfgang Löffler. With the kind permission of Wolfgang Löffler)

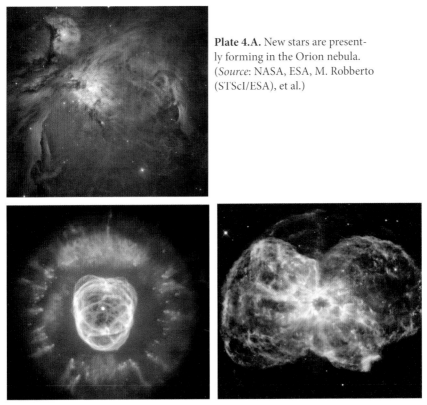

Plate 4.A. New stars are present-
ly forming in the Orion nebula.
(*Source*: NASA, ESA, M. Robberto
(STScI/ESA), et al.)

Plate 4.B. Planetary nebulae come in many beautiful colors and impressive shapes. *Left*:
The Clownface Nebula (NGC 2392). *Right*: Planetary Nebula NGC 2440. (*Sources: Left*,
Andrew Fruchter, STScI et al. WFPC2, HST, NASA. *Right*, NASA, ESA, K. Noll, STScI.
Hubble Heritage Team, STScI/AURA)

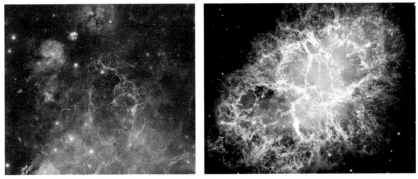

Plate 4.C. *Left*: The heavy-element-enriched supernova remnant Vela is the result of a
Type II supernova explosion. *Right*: The Crab Nebula (sometimes also called the Crab)
is likewise the remnant of an exploded massive star. (*Sources: Left*, copyright Davide De
Martin. *Right*, NASA, ESA and Allison Loll/Jeff Hester, Arizona State University. Davide
De Martin, ESA/Hubble)

Plate 6.A. The Milky Way as seen from the Southern Hemisphere. (*Source*: Printed with kind permission of Gabor Furesz)

Plate 6.B. The Magellan Telescopes (*left*: "Clay," *right*: "Baade") under the starlight of the Milky Way. The Magellanic Clouds are clearly discernible at the upper left. (*Source*: Printed with kind permission of Gabor Furesz)

Plate 6.C. The spiral galaxy NGC 6744. The Milky Way would likely look like this galaxy if we could see it from the outside. (*Source*: ESO)

Plate 6.D. Illustration of the spiral structure of the Milky Way and our position (at "Sun") in the Orion-Cygnus Arm ("Orion Spur"). (*Source*: R. Hurt (SSC), JPL-Caltech, NASA. Survey credit: GLIMPSE)

Plate 6.E. The author in front of the Magellan-Clay Telescope during a longer exposure on one of the dwarf galaxy stars. The Milky Way is clearly recognizable in the star-studded sky. (*Source*: Printed with kind permission of Gabor Furesz)

Plate 6.F. The Pleiades, also called Seven Sisters, in the Taurus constellation. It is a young open star cluster whose brightest stars are easily visible to the naked eye. (*Source*: NASA,ESA, and AURA/Caltech)

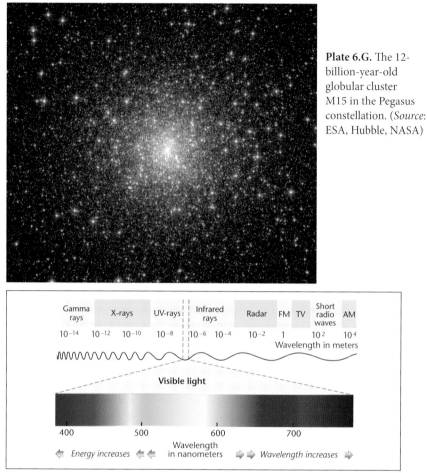

Plate 6.G. The 12-billion-year-old globular cluster M15 in the Pegasus constellation. (*Source*: ESA, Hubble, NASA)

Plate 7.A. The visible light spectrum with its rainbow colors. It is just a small segment of the complete electromagnetic spectrum. Various wavelength ranges are indicated. (*Source*: Peter Palm)

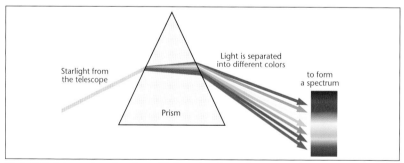

Plate 7.B. Schematic refraction of light going through a prism. A spectrograph splits up starlight into a spectrum in the same way. The resulting spectrum can then be processed and analyzed. (*Source*: Peter Palm)

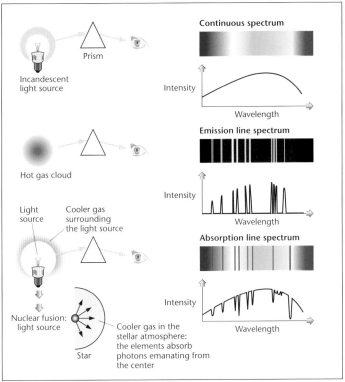

Plate 7.C. Schematic view of the origin of spectral lines. *Top*: Continuous spectrum of an incandescent light source. *Middle*: Emission line spectrum of a hot gas. *Bottom*: Absorption line spectrum of a star. (*Source*: Peter Palm)

Plate 7.D. At sunset, the two 6.5-m Magellan Telescopes are prepared for observations. "Baade" is on the left, "Clay" on the right. (*Source*: Anna Frebel)

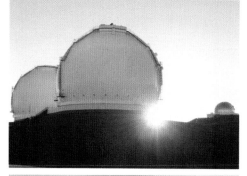

Plate 7.E. *Top left*: Above the clouds at 4,000 m on the summit of Mauna Kea, Hawai'i, USA. *Top right*: The Japanese 8-m Subaru Telescope with the control building next to it. *Bottom*: The neighboring 10-m Keck Telescopes during sunrise. (*Source*: Anna Frebel)

Plate 7.F. *Top*: Control room of the Clay Telescope with the observer (the author) and the telescope operator. *Bottom*: Computer screen displaying the user interface of the spectrograph (left), a newly taken stellar spectrum (upper right), and a preliminary data analysis to verify the proper exposure time (lower right). (*Source*: Anna Frebel)

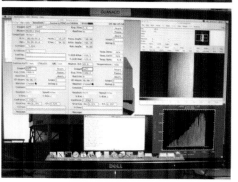

Plate 8.A. *Top*: The welcome sign upon entering the observatory that is operated by the Australian National University. *Bottom*: Siding Spring Observatory hosts an entire population of telescopes. The boxy 2.3-m telescope with its external staircase can be seen in the foreground. (*Source*: *Top*, Anna Frebel. *Bottom*, printed with the kind permission of the Australian National University)

Plate 8.B. *Top*: The 2.3-m telescope in its rectangular, rotating building during sunset and during the day. *Middle*: The control room with quite a few computer screens and sweets to stay awake. *Bottom*: Two excited but tired observers. (*Source*: *Top*, printed with the kind permission of Harvey Butcher, Australian National University. *Center and Bottom*: Anna Frebel)

Plate 8.C. Observing with the 2.3-m telescope. *Top*: To observe a star, the telescope points through the large opening in the dome. Very bright stars are already observable at dusk. *Lower left*: The telescope in its park position. The mirror is protected by the large covers surrounding the black tube. The light first falls onto the primary mirror and then onto the small secondary mirror above. From there, it is directed through the black tube and from there to the instruments. *Lower right*: The "double-beam spectrograph" is mounted on the telescope at the bottom right at about the height of the mirror. It splits the light into red and blue beams, which are then simultaneously sent through two spectrographs. (*Source*: Anna Frebel)

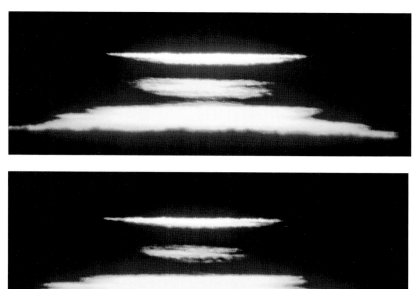

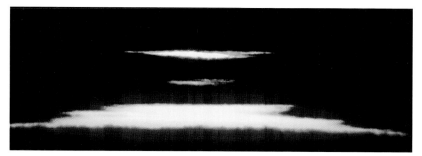

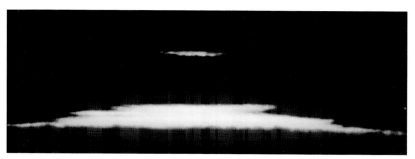

Plate 8.D. The evolution of a "green flash" at the upper rim of the solar disk at sunset. During this event, the upper end of the solar disk turns dark green for just a split second. (*Source*: Printed with kind permission of Gabor Furesz)

Plate 8.E. A dramatic sunset in Chile. Clouds make for an impressive twilight. Obviously, clouds are undesirable for observations. (*Source*: Anna Frebel)

Plate 10.A. *Top left*: View from the car not far from where the bushfire had, shortly before, entered Canberra on January 18, 2003. *Top right and bottom*: Many houses were quickly burned down to ground. Burned-out cars and piles of bricks were all that was left behind. (*Source*: Anna Frebel)

Plate 10.B. The depressing view of burned-down telescopes at Mt. Stromlo Observatory. Its visitor center complete with park benches had been a favorite stop for residents and tourists alike. *Top and middle*: Building of the Columbia Telescope before and after. *Bottom*: The burned refractor inside the Columbia Telescope. *(Source:* Anna Frebel)

Plate 10.C. The Commonwealth Solar Observatory building before the bushfire in January 2003 *(top)*, directly after the fire *(middle)* and after its reconstruction *(bottom)*. Close attention was devoted to restoring the building to its original design. (*Source*: Anna Frebel)

Plate 10.D. Burned bookshelves in the ruins of the Commonwealth Solar Observatory library. (*Source*: Anna Frebel)

Plate 11.A. The 1.3-m SkyMapper Telescope at Siding Spring Observatory at dusk. (*Source*: Research School of Astronomy and Astrophysics, Australian National University. Printed with kind permission of Martyn Pearce)

Plate 11.B. An artist's rendering of the planned Giant Magellan Telescope with its huge segmented mirror, 25 m in diameter. Observations will commence in the early 2020s. Note the truck for size comparison. (*Source*: Printed with kind permission of Giant Magellan Telescope—GMTO Corporation)

In my search for the most metal-poor stars, the discovery of HE 1523-0901, one of the oldest and best dated stars, was particularly exciting. I had given a few of my brighter stars to an astronomer friend of mine to fill in a gap in his target list while observing with the 6.5-m Magellan Telescope in Chile. Shortly afterward, I left with my thesis advisor for the 4-m Australian Astronomical Telescope at Siding Spring Observatory in Australia, to search for r-process stars in my sample myself. On one of those April nights, I received an email from my friend with the first processed data. While one of my longer exposures was still in progress, I could not keep myself from immediately inspecting the new spectra. Whoops! What's this? The huge europium line made my jaw drop. Since such strong europium lines can also occur in very metal-rich stars as a result of billions of years of chemical evolution, I nervously decided to quickly carry out a few tests to verify that my candidate was indeed a metal-poor star. Everything was okay on that front, and I became quite excited. My advisor and I soon inspected other regions in the spectrum to assure ourselves that we really were dealing with an exciting r-process star.

Then, one thing became clear: There we were, sitting in a telescope, looking for r-process stars, and precisely one of those just happened to come in by email. Was it chance? Fate? Providence? Whatever it was, we decided to observe the star straightaway to obtain better data. Alas, the weather was less on board with our plan. It was pretty hazy that night, and all of our attempts to gather data better than the Magellan spectrum from my friend were completely in vain. That is simply how it sometimes goes in a field so dependent on the weather. The only thing you can do is take a few deep breaths—being too annoyed will not really help, unfortunately.

In the very end, the bad weather was not a prohibitive obstacle, though. Back home again in Canberra, I had time to prepare a preliminary analysis of the Magellan spectrum. With those results, I wanted to apply for immediate telescope time at ESO's Very Large Telescope. In very special cases it is possible to get a small amount of telescope time at short notice, without having to wait for an entire year before the object becomes observable again. This application was indeed granted. Almost three months had already passed since that first observation, and HE 1523-0901 was observable for only a short while longer, until mid-August. Half of the observations applied for were executed in so-called service mode before the star completely vanished below the horizon for about half a year. Then, I had to write yet another proposal to get the other half of the required telescope time.

In keeping with the saying "good things take time," late that following summer I received the full set of data on a DVD in the mail (because the data were taken in service mode). It was almost one and a half years after the star had first been observed by my friend. At that time, I was in Uppsala, Sweden, for several weeks to work with colleagues at the university's astronomy department. That very week, at Friday tea, I gave a first informal presentation on my preliminary results of the new spectrum. I was excited for my coworkers to be the first ones to hear about this new discovery.

As fate would have it, another star that my friend had observed with the Magellan Telescope at the same time as HE 1523-0901 was also quite unusual and interesting: a carbon-rich s-process star that in addition also exhibited signs of r-process enrichment. This star had been forgotten for a while, mostly over the excitement of HE 1523-0901. When I "found" it again on my computer I had to laugh out loud. It seemed too good to be true. I then handed over the data to one of my students so that she could adopt the star and carry out an abundance analysis and interpretation.

5.4 Nuclear Astrophysics

R-process stars connect astrophysics with nuclear physics in a unique way. Owing to their old age, these objects are of cosmological significance to astrophysics. At the same time, they act as cosmic laboratories to nuclear physics because metal-poor stars are the bearers of the chemical fingerprints of the various processes of nucleosynthesis that provide crucial experimental data on the synthesis of the heaviest elements in the Universe. The quest to study the r-process and to determine stellar ages conjoins these two areas of research because the ages directly depend on our nuclear physics knowledge of the structure and behavior of atomic nuclei. For this reason, astronomers, theoretical nuclear physicists, and experimentalists work together to derive as much information as possible on the production mechanisms and site(s) from the element abundances of r-process stars but also s-process stars, in the field of nuclear astrophysics.

In this context, lead serves again as a good example. Although the lead in the Universe is mainly synthesized by the s-process at later epochs, some lead production had already occurred in the r-process during the earliest times. On one hand, lead is generated directly during the r-process through β-decay of the heaviest, most neutron-rich isotopes, including those in the transuranium region. On the other hand, it builds up on cosmic timescales through the slow α-decay of thorium and uranium.

Lead in an r-process star is even more difficult to measure than uranium. The sole lead absorption line, at 405.8 nm, is even weaker than the already weak uranium line, and thus requires even higher-quality data. Still, the lead line could be detected in CS 31082-001 and HE 1523-0901. As different r-process models do indeed predict different abundance distributions for the very heaviest elements, tests for self-consistency of each model and for agreement with observations are particularly important and informative. A consistency test can be carried out if the abundances of thorium, uranium, and lead can all be measured in one and the same star since the abundances of those three elements are linked to each other due to their production in the r-process and through radioactive decay. The challenge lies in predicting the various contributions to the lead production. The thorium and uranium abundances left over after 13 billion years of decay have to be correctly forecast at the same time. Results from the comparison with the observational data can in turn be used to improve the next generation of r-process models. This will ensure better predictions of r-process abundances in the supernova (or whatever the production site may be), which will ultimately also yield more accurate stellar ages with lower uncertainties.

CHAPTER 6

WELCOME TO OUR MILKY WAY

Our home Galaxy, the Milky Way, may seem inconceivably huge on the sky when viewed at night. Its size easily exceeds one's power of imagination. But when you look more closely, it turns out that even such a huge galaxy has a distinct structure of its own, making it actually seem not quite so huge anymore. It is composed of gas and dust, millions of individual stars, large star clusters, and even dwarf galaxies. Next, our Galaxy can be divided into various components, each of which has its own story to tell. We will thus examine the individual Galactic regions with their various cosmic inhabitants in more detail.

6.1 A Milky Way above Us

Anyone who has ever seen the Milky Way from the Northern Hemisphere has probably been as impressed as I was. It was always thrilling to imagine how many stars are shining out there. Once in Australia, on a cold but extremely clear winter night, I finally saw the Milky Way in its full glory for the very first time. In that moment, it became clear to me that astronomy really was "my thing" and something that I had to pursue. As can be seen in Plate 6.A, the Milky Way does indeed look considerably more dramatic in the Southern Hemisphere because it shows more stars and more structure. It tremendously motivated me to become an astronomer and best of all, right there, in Australia.

Some years later, I was operating one of the large telescopes in Chile, the 6.5-m Magellan Clay Telescope, as part of my ongoing research on dwarf galaxy stars. On a clear, starry night some of my coworkers had found a small, fully automatic amateur telescope in a tucked-away storage room and decided to try it out. These coworkers were building instruments for large telescopes and had flown to Chile solely to install a new instrument on one of the two Magellan Telescopes. Now they wanted to have some fun with the little telescope as they did not have any observing to do during the night. They called me at "my" telescope to inform me that they were stargazing and that I should drop by if my observations would allow it. Soon after, I decided to walk halfway down the mountain ridge during a long, 55-minute exposure of one of my dwarf galaxy stars. I, too, wanted to take a direct look into the cosmos with my own eyes. My professional computer-driven Magellan Telescope could do very well without me for a short while under the supervision of the telescope operator.

I knew that a narrow path with some wooden steps led directly from the telescope down the mountain ridge. It had been quite a while since I had last followed the path, though. The alternative was the paved road that led further out around the mountaintop and was less steep. I was a little unsure about whether I would find the narrow path quickly enough in the dark, but because I had only 55 minutes to spare, it would clearly be the faster route down and back up. Driving by car was not really an option either because even just the parking and brake lights while driving away could cause significant light pollution to the exposure of my precious observation, which I did not want to risk.

So, off I went. It was about 10 p.m. and a late-summer night in the Andes at an altitude of 2,400 m. Despite a light breeze, it was still relatively warm. The Moon was not visible anywhere. It was pitch dark and all I could see was darkness. As I was passing, a pair of red- and orange-colored lights blinked at me from the telescope buildings, like small red eyes. It took me a few minutes to adjust to the dark. But with the help of my flashlight I soon found the right path.

The small, nifty telescope turned out to be a pretty interesting "toy." It whirred from object to the next object at the push of a button. Several tiny operating lights were blinking in different colors while observing. All that was needed was a tinny voice, like that of R2D2—that little robot from Star Wars—to provide a commentary. Such telescope programs probably already exist, but our little telescope stayed mute. From the menu that was attached to its side we could choose which of the many programmed objects we wanted to see. A pretty globular cluster, perhaps? A colorful planetary nebula? Or a galaxy, after all? There are quite a lot of interesting things to see—even with a small telescope.

All of us felt like genuine "stargazers," and we had fun pointing the telescope from one object in the sky to the next, over and over again. Rrrrrsssst, rrrrrrrssst, and then one more time, and again, and yet again. Then, my time was up and I had to return to my telescope to inspect my new spectrum and start another exposure. Since I had safely followed the little path down, I also wanted to take it back up again.

Apart from the occasional brief on-and-off switching of our flashlights, it had been pitch dark throughout the stargazing with my co-workers. My eyes were thus very well adjusted to the dark. Still, I embarked with my flashlight in hand to find the path. I noticed right away, though, that I did not need the flashlight after all. I stopped to take a deep breath and looked directly above toward the sky. The Milky Way was shining mightily above me. It was so bright that I could see the path ahead of me zigzagging between some low shrubbery and plenty of rocks and gravel. The Moon was still nowhere to be seen. And yet, I was astonished to realize that I really could see enough to find my way back to the telescope entirely without flashlight. A short while later, I reached the control room having been guided solely by the light of the stars. Plate 6.B depicts the Magellan "Clay" and "Baade" Telescopes faintly illuminated by the Milky Way and the last red rays of sunset on the horizon. Since our eyes cannot perform as long an exposure as a camera can, such photos of course appear somewhat brighter than what I could see that night.

On my walk back, I recalled some stories I had heard while in Australia, about aborigines in the Australian outback. Over the millennia, they had been wandering about at night—without any light source, of course, except for the Moon, which certainly can be very bright. But they probably did not need any. They had the Milky Way overhead. The ancient Egyptians also oriented their daily routines by the stars, and all the early seafarers were completely dependent on stars for navigation. At that moment, I knew from my own experience that in a region without light and air pollution, starlight does indeed suffice for moving around at night without any extra light, on foot at least. Ever since then, I have been more careful about contributing as little as possible to light pollution by simply switching off unnecessary lamps. It saves energy as well. Those popular "Earth at Night" posters do look fantastic because you can see all the major cities and inhabited continents lit up by night. When you think about it, though, this much light poses a growing problem. Today, preserving the dark night sky seems to have low priority. Fortunately, some organizations are drawing attention to this problem and its numerous negative consequences, for instance, for the flight paths of migrating birds. Astronomers and astronomy enthusiasts are not the only ones affected by this issue.

Travelers to Australia, South Africa, and South America should definitely add "stargazing" to their list of attractions. This free-for-all show can be enjoyed on a clear night away from any city. But it has to be dark, of course—only a few places on Earth are still free of light pollution these days. Relatively dark areas can be reached quickly, though, perhaps after a 20- to 30-minute drive from the city, up a mountain, or even along the coast. It should also be kept in mind when planning such a trip that the Milky Way is positioned directly overhead in the Southern Hemisphere around midyear, that is, in wintertime. It is also the prettiest then. At other times of the year, it is still visible near the horizon but then only shortly after dusk or before dawn.

In our very hectic world today, we should more often remember what kind of spectacle is happening above us and should occasionally behold it. With some luck, you might have the additional reward of a fleeting glimpse of a shooting star or two. Then there is hope that your wish will come true . . . right? Who would not want to try this sometime!

6.2 The Milky Way's Structure

We live inside the Milky Way. This means that our view into the cosmos is unfortunately somewhat limited. Thus, creativity is needed to find out what the Galaxy might actually look like and what its structure is. Owing to its enormous size, we will never be able to view it from the outside as a whole like we see other galaxies from afar. We are like goldfish trying to discover whether our aquarium is standing in a garage or on the 12th floor of a high-rise building. Fortunately, various observations of stars and other galaxies, such as the Andromeda galaxy, provide some answers to this important question.

The ancient Greeks and many native peoples before them began to study the night sky with the simplest of optical instruments: the human eye. What we can see today—leaving aside light pollution and bright moonshine—is that the stars are mainly assembled in a broad, diffuse band across the sky.

The name "Milky Way" goes back to the ancient Greeks. We can read how the Milky Way was formed in classical mythology: In yet another one of his infidelities, the notorious womanizer and king of the gods Zeus had sired a son, Heracles (whom the Romans called Hercules). His wife, the goddess Hera, was understandably filled with rage and jealousy. As Heracles's mother was a mortal, he was not immortal like the gods, so Zeus resorted to some craftiness to remedy this shortcoming. The breast milk of the sleeping Hera was supposed to help baby Heracles gain immortality, so he was secretly placed with her. But the infant nursed so energetically that Hera woke up. All angry, she flung the boy away from her breast. Her milk then sprayed all the way up to the sky, where it can still be seen as the Milky Way. A nice little myth—but what is the Milky Way in reality?

In 1610, the Italian scientist and inventor Galileo Galilei was the first person to examine the Milky Way with a telescope, which had only recently been invented. He found that the shimmering band across the sky consists of innumerable faint stars. In the 19th century, the English astronomer William Herschel was also interested in the night sky and its stars and published the first map of what he thought the Milky Way might look like from the outside. Today we know that most of the stars of the Milky Way are arranged inside a flat discus-like plate that provides the basic shape of our Galaxy. But why do stars arrange themselves in this particular manner?

Around 1900, astronomers began to examine the Milky Way's structure by stellar statistical methods, which proved difficult. Henrietta Leavitt's period-luminosity relation finally provided the right means for progress. For stars that periodically vary in brightness, for example the so-called Cepheid variables, this relation describes how the period of the variable light relates to the star's luminosity. Around 1912, Harlow Shapley applied this relation for the first time to perform distance measurements of Cepheids. Around 1920, distances to cosmic structures inside and outside the Milky Way were measurable for the first time with this technique, as long as Cepheids could be found there.

Around 1927, Jan Oort and other scientists largely succeeded in explaining the motions of stars in the Milky Way with geometry and a few targeted observations. They found that stars revolve around the Galactic center under the influence of the gravitation caused by a strongly concentrated mass at the center. The stars are arranged in a disk and behave like a fluid that rotates more quickly at the center than farther out.

It was soon recognized that the Galaxy has a detailed spiral structure, since the stars, gas, and dust of the disk are all arranged in a number of spiral arms. The subsequent decades produced countless observations that once and for all confirmed the Milky Way's disklike shape. Radio-astronomy can detect hydrogen gas, which is what the spiral arms are mainly composed of. Over the decades, it became possible to map the spiral structure of the Milky Way in great detail this way. Consequently, our home Galaxy is classified as a spiral galaxy. It is now assumed that the Milky Way very much resembles the Andromeda galaxy and particularly the galaxy NGC 6744. Both are depicted in Plate 6.C.

If you want to imagine the Milky Way as whole, you can think of it as a quite thick and fluffy pancake with generous layers of jam and cream spread on top and underneath, crowned by a centrally positioned scoop of ice cream. Finally, add a cherry on top of the ice cream. The jam and cream symbolize different stellar populations distributed over the disk that add to the pancake's thickness. The scoop of ice cream corresponds to our Galaxy's "bulge." The bulge is a large dense assembly of countless stars at the center and thus is the most luminous part of the Galaxy. Observations indicate that the spiral arms in the inner part of the Milky Way merge into a bar-shaped structure. Finally, at the very center of the bulge there is a supermassive black hole of 4 million solar masses—the cherry. This monster devours vast quantities of stars and gas in the inner parts of the bulge.

The disk including its spiral arms has a diameter of over 100,000 light-years and is roughly 1,000 light-years thick—a length of 1,000 light-years corresponds to 9.5 quadrillion km. Where are we located with respect to the center of the Galaxy? Luckily, the Solar System is relatively far out in the disk so that the black hole constitutes no danger. Our location is 28,000 light-years away, about two-thirds of the distance between the center and the edge. This is illustrated in Plate 6.D. Whereas the Earth is revolving around the Sun at 30 km/s, the Sun is simultaneously moving with the Solar System at about 220 km/s along a slightly elliptic orbit within its spiral arm around the Galactic center and the bulge. One such revolution lasts about 250 million years. The Sun, with its 4.6 billion years of age, has accordingly circled around the center about 20 times. The evolutionary stage leading to higher life forms on Earth, such as mammals, over the past some 200 million years does not even correspond to one "Galactic year."

Our exact position in the Milky Way is in the local spiral arm that is also referred to as the Orion-Cygnus Arm. The Sun and the Solar System lie at the inner edge of this arm. There are four larger and two smaller arms in total, which are represented schematically in Plate 6.D. The spiral arm laying in the direction of the center, seen from us, is the Sagittarius Arm, and the one "behind" us is called the Perseus Arm.

Since we reside inside a spiral arm, we can see a portion of three spiral arms across the sky—stars from our own arm as well as stars and gas from both neighboring arms. It is the spiral structure of the Galactic disk that leads to the bandlike character of the "Milky Way" on our night sky: from the Northern Hemisphere, we see directly into the Perseus Arm behind us. But, from the Southern Hemisphere, the Milky Way looks even more magnificent because it is the view into the Sagittarius Arm in the direction of the center of our home Galaxy. A lot is going on there, from the astronomical point of view. The countless stars in the region of the bulge provide the milky background light that makes the Southern Hemisphere Milky Way seem even brighter and prettier than what just the Sagittarius Arm alone would look like.

The Milky Way divides the sky into two parts, roughly equal in size. From this, we can deduce that the Sun and the Solar System are located in the main plane of the disk. Hence, we are sitting in the vertical middle of the pancake. When we look up or down, we can see outside of the pancake but if we look around ourselves along the pancake plane, we see nothing but "pancake"—that is, the Milky Way band around us.

Taking a closer look at the Milky Way on the sky, you will quickly see that it is not homogeneously lit up but exhibits very many smaller and larger structures. Dark regions alternate with lighter ones, and the number of visible stars can be very variable. The darker spots—a particularly prominent one is called the "Coalsack"—are regions where starlight on its way to us is completely blocked by very dense interstellar dust and gas clouds. "Dust" refers to tiny particles such as dust grains or even small aggregates clumped together to form larger particles composed of various elements, such as carbon or silicon. In the Galactic disk, huge

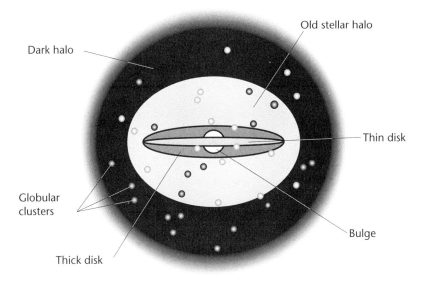

Figure 6.1. Edge-on view of the Milky Way with the different components of the Galaxy indicated. (*Source*: Peter Palm)

quantities of dust are present, especially at sites of star formation. Dust assists in the formation of stars by cooling the gas, enabling it to clump and condense into stars. The Galactic center is also hidden from us behind these dark dust clouds and is accessible to direct observation only by means of radio astronomy. Thus, the dark regions in the band of the Milky Way are not caused by dark matter, clouds in the Earth's atmosphere, or even effects of the eye, contrary to any first impression when looking at the night sky.

The disk of the Milky Way is densely populated by young stars—95% of all Galactic stars are found there. It consists of the larger, so-called thin disk, which itself is surrounded by the "thick disk." Figure 6.1 shows an edge-on view of the Milky Way, illustrating the different disks. The disk components are embedded in a large spherically shaped region that is referred to as the stellar halo. The halo has a considerably lower concentration of stars than the disk and primarily contains older stars as well as star clusters and some dwarf galaxies. This stellar population is especially important for stellar archaeology because the oldest and most metal-poor stars reside there. All halo objects revolve around the center of the Milky Way on large, fairly circular orbits. The halo extends over many hundred thousands of light-years around the Milky Way.

The number of stars present in the Milky Way is difficult to estimate because the transition from the galaxy to interstellar space is smooth and hard to identify. Where our pancake ends is very clear. A galaxy has no sharp edge though. The distribution of stars in a galaxy can be simulated with computer models, but various assumptions, for example on how stellar concentration varies with increasing distance from the Galactic center, quickly lead to divergent results. The various stellar populations contribute differently to the stellar concentration in different parts of the Galaxy. The concentration of stars in the halo, for instance, is considerably lower than in the disk. Finally, the different types of stars are also important. How many of each type and class of stars are in the Galaxy has to be estimated. This giant jigsaw puzzle reveals that there are between 200 and 400 billion stars in our home Galaxy, most of which are found in the disk and the dense central bulge region. The Sun is merely one of these stars. Compared to our sister galaxy Andromeda, however, the Milky Way is a smaller fish in the Universe. The Andromeda galaxy has roughly one trillion stars, therefore, three to five times as many.

Finally, the whole Milky Way Galaxy is embedded inside a halo of dark matter. This dark halo is much larger than the stellar halo, although it is inaccessible to direct observation and can be studied only indirectly. In fact, every galaxy is surrounded by a dark halo that holds it together gravitationally. Consequently, the luminous part of the Milky Way is just a small part of the actual galaxy, which, in the end, mostly comprises dark matter. Counting the dark matter, the Milky Way has a mass of about one trillion solar masses, which corresponds to 10^{42} kg (it is a one followed by 42 zeros).

The only way to detect dark matter is by its gravitation. Dark matter in and around a galaxy reveals itself in what is called rotation-curve analysis. In such an analysis the rotational velocity of a galaxy is measured by means of the Doppler effect at different distances away from galaxy's center. Based on the star distribution within the galaxy, it would be expected that galaxies rotate more quickly near their center than farther out. There, it would rather be slower, just as when water runs in a spiral down the drain. Very far out, it would hardly rotate at all. Various observations of spiral galaxies have found, however, that there still is a considerable measurable rotation in the outer regions. Judging from the luminous material alone, this cannot be possible. These observations are, however, readily explained if you assume the existence of additional, yet invisible, dark matter that is predominantly present in the outer parts of the galaxy and is rotating along with the luminous matter.

Astronomers have been working with dark matter for many decades by now, even though no interactions between it and luminous matter have been observed. Simulations of the evolution of dark matter in the Universe are of great importance to our understanding of structure formation, and of the formation of galaxies and their evolution. Yet neither astronomers nor physicists know what dark matter, in fact, consists of. Many experiments are currently being conducted in hopes of detecting dark matter directly as an elementary particle. One possible type of dark matter particle, some kind of a weakly interacting massive particles (WIMP), may be the most promising, but we shall have to wait and see what the solution will be.

6.3 Dwarf Galaxies

The Milky Way is not alone in this region of the Universe. It is part of the so-called Local Group, together with its sister spiral galaxy in the Andromeda constellation, the somewhat fainter Triangulum galaxy (another spiral galaxy), and more than 60 other smaller galaxies. Figure 6.2 shows the Local Group's arrangement in space. This group is held together by the mutual gravitational attraction and forms a kind of family of galaxies. This family is part of the Virgo Supercluster, which, in turn, is composed of some large galaxy clusters and other galaxy families.

After the three main galaxies, the Magellanic Clouds are next in mass and size. These two galaxies are situated in the neighborhood of the Milky Way and are generally regarded as the largest examples of so-called dwarf galaxies. They can be seen by the unaided eye as moderately bright, diffuse spots ("clouds") in the Southern Hemisphere sky. They owe their name to the seafarer Ferdinand Magellan, who discovered them for the Spanish king during the 16th century and navigated by their light at night. As the term indicates, dwarf galaxies are substantially smaller than the Milky Way and have a mass of less than 100 million solar masses.

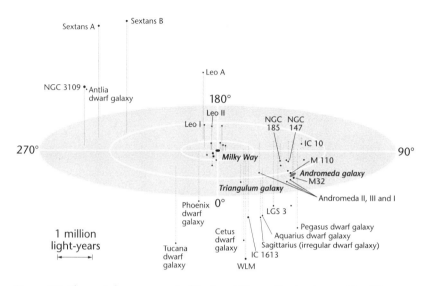

Figure 6.2. The spatial arrangement of the Local Group. It extends over 10 million light-years and consists of the Milky Way, Andromeda, and the Triangulum galaxy as well as many smaller dwarf galaxies. (*Source*: Peter Palm)

Many dwarf galaxies in the Local Group are located in the stellar halo of the Milky Way. For this reason they are often called satellites or satellite galaxies. The Milky Way has about 30 such companions that are gravitationally bound to it. The Andromeda galaxy has considerably more of them. However, some dwarf galaxies are not bound to any larger galaxy, neither to the Milky Way nor to Andromeda but are still part of the Local Group

But not all dwarf galaxies are alike. A whole range of dwarf galaxies with different properties have been found so far. Accordingly, dwarf galaxies can be divided into several types: irregular, spheroidal, and elliptical. Dwarf irregular galaxies are relatively young, just a few billion years old, and have a lot of gas so they can continue to form new stars. Compared to them, dwarf spheroidal galaxies are old and deficient in gas. All stars in them formed a long time ago, before the galaxy had exhausted its gas supply. Dwarf elliptical galaxies are rather elongated in shape and still have enough gas available for star formation. All these dwarf galaxies (often just briefly referred to as dwarfs) are galaxies in their own right, though. Composed of luminous matter, that is, stars and gas, they are surrounded by halos of dark matter. This property distinguishes dwarf galaxies from star clusters, which lack a dark halo (such as globular clusters).

Although any relationship between these types of galaxies is still largely unresolved, detailed studies of the entire population of dwarf galaxies in the Local Group indicate how star formation proceeded in the various subgroups and what characteristics their interstellar medium has. This, in turn, provides information for a better understanding of the origin of the Local Group and large spiral galaxies.

Dwarf spheroidal galaxies from the Local Group have a broad range of luminosities. Interestingly, the luminosities of the faintest dwarf galaxies known today are dominated by their own brightest stars. Thus, their luminosity is only a few thousand times larger than that of the Sun. But the Sun is not a particularly bright star. By contrast, other dwarf spheroidal galaxies have luminosities up to 20 million times the Sun's luminosity. The Magellanic Clouds are even more luminous than that because they belong to the group of dwarf irregular galaxies and contain many more stars and significant amounts of gas.

All these luminosities are still quite low compared to that of the Andromeda galaxy. This spiral galaxy has a luminosity of 10 billion solar luminosities. It is now clear where dwarf galaxies get their name from: they are the tiny fireflies of the Universe. But since they can be found everywhere, they should not be underrated. Their large numbers contribute significantly to the total luminosity of galaxy groups, such as in the Local Group or even larger clusters of galaxies.

We will now be dealing mainly with ancient dwarf spheroidal galaxies because they are the ones containing old metal-poor stars. This suddenly makes an entire class of galaxies of interest to stellar archaeology. Harlow Shapley discovered two of these dwarf spheroidal galaxies as early as 1938: Fornax, that is, the "chemical furnace," and Sculptor, named after the constellations in which they are located. Figure 6.3 depicts the Sculptor dwarf galaxy. Interestingly, Fornax is orbited by a number of globular star clusters showcasing that globular clusters can form under a variety of conditions. Leo I and Leo II, as well as Draco and Ursa Minor, were discovered later, in the 1950s. Finally, Carina was added in 1977 and Sextans in 1990. Since they comprise hundreds of thousands of stars and were discovered some decades ago, these

Figure 6.3. The Sculptor dwarf galaxy in the Sculptor constellation. It is one of the more luminous "classical" dwarf galaxies. (*Source*: "Sculptor Dwarf Galaxy." Anglo-Australian Observatory and the Royal Observatory of Edinburgh, for Digitized Sky Survey images)

satellites are now often referred to as the "classical" dwarf galaxies. They possess total luminosities between 200,000 (Draco) and 20 million (Fornax) solar luminosities.

One great advantage of these satellites, along with most dwarfs in the Local Group, is that they are not too far away from us compared to other galaxies: merely between 200,000 and 800,000 light-years. Andromeda, for example, is 2.5 million light-years away. This means that detailed observations of these dwarfs is just possible. Individual stars can be resolved in an image, for instance, whereas fainter galaxies appear merely as a diffuse blob. Spectroscopic observation of the satellites' stars are a challenge, though, as they are almost too faint. Only the brightest stars in these galaxies are just observable with the largest telescopes.

Dwarf spheroidal galaxies are relatively simple systems—with an emphasis on "relatively." Compared to the gas-rich dwarfs, they seem to be quite uncomplicated. However, countless studies over the past five decades conclude that they are very ancient systems that have undergone many early episodes of star formation. Furthermore, their slower chemical evolution makes them generally metal-poor galaxies. Slow evolution means that these galaxies spent their gas on star formation before they could produce metal-rich stars in large quantities in the way other galaxies did. Therefore, the average metallicity of these galaxies is lower than, say, irregular systems, which still have a lot of gas at their disposal for continued star formation.

This is the reason why all the dwarf galaxies obey a metallicity-luminosity relation. The more luminous dwarfs have higher metallicities than fainter galaxies. In stellar archaeology, this means that particularly the darkest, smallest dwarfs are potentially the most interesting systems. The concentration of old, metal-poor stars is the highest there.

Since 2005, more than 15 additional satellite galaxies have been discovered. They are extremely faint, and could be found only in new, deep, large-scale sky survey images. Due to their low luminosities these galaxies appear, at best, as barely noticeable stellar overdensities in a particular patch of sky. Even then, usually no more than a few dozen of the brightest stars can be identified as actual members. Overall, those galaxies cannot really be "seen" anymore. Sophisticated computer programs with special search algorithms are necessary to pick out member stars on the sky amid the numerous foreground stars. The luminosities of those galaxies are 10 to 100 times less than the luminosities of the classical dwarfs.

The number of known satellites in the vicinity of the Milky Way suddenly doubled with the discovery of ultra-faint "minigalaxies." The Sloan Digital Sky Survey can be credited for this accomplishment. It recorded one-fourth of the Northern Hemisphere sky with the wide-angle camera of the 2.5-m telescope at Apache Point in New Mexico. Other systematic surveys, mainly in the Southern Hemisphere, are in the process of discovering more of these cosmic fireflies.

Does this mean that, in accordance with the metallicity-luminosity relation, these new ultra-faint dwarf galaxies are all very deficient in metals? The answer is definitely yes. Some studies have shown that these dwarfs are mainly composed of metal-poor stars and contain no stars with high, solarlike metallicities. Hence, these minigalaxies are a gold mine for stellar archaeologists. Indeed, several extremely metal-deficient stars have been discovered over the past three years. All in all, about 30% of all known extremely metal-poor stars have since been found to be in dwarf galaxies. My colleagues and I observed many of these stars with the Magellan Clay Telescope in Chile and the Keck Telescope on Hawai'i. Our aim was to determine their metallicities and detailed chemical element abundances. I was observing one of these precious stars during that night of my walk under the Milky Way in Chile. Plate 6.E shows the telescope and the beautiful night sky during one of those observations.

It is remarkable that the ultra-faint dwarf galaxies have particularly high concentrations of dark matter. This has turned them into popular objects for dark matter detection experiments in the hope of directly detecting dark matter signals someday. Indeed, they have already provided interesting results about the distribution of dark matter in galaxies and their characteristics.

Large sky surveys such as the Sloan Digital Sky Survey have led to discoveries of not just new dwarf galaxies but also quite a few giant stellar streams distributed across the sky (see Plate 1.D). These thin bands of stars stretch across major sections of the sky. They likely originate from disintegrated and still disintegrating dwarf galaxies that came to die in the Milky Way's gravitational field. This way, many dwarf galaxies have been "devoured" by our Galaxy during its assembly over the course of billions of years. For small objects in the gravitational field of a larger galaxy, such fate is very common.

Slowly but surely, the Milky Way will continue to incorporate more dwarf galaxies into its halo in the future. On the basis of detailed observations of stars in these streams, it is possible to reconstruct what kind of dwarf galaxy had been torn apart and roughly how large it was. These findings help us understand how galaxy accretion events influenced the evolutionary history of the Milky Way and what has and still is contributing toward the buildup of the Galactic halo. This process also shows that the evolution of a large galaxy such as the Milky Way never will really come to an end.

6.4 Star Clusters

Most of the stars we see on the night sky seem to stand there just by themselves. Yet, many stars are also arranged in tight groups that are visible as such on the sky. These groups are called star clusters. Depending on the number of stars, open star clusters and globular clusters are distinguished. The Pleiades in the Taurus constellation, also known as the Seven Sisters, is a prominent example. It is a young, open star cluster about 115 million years old whose six to nine brightest stars are easily discernible with the naked eye. They are shown in Plate 6.F. An open star cluster has just a few hundred thousand members, at most, whereas a globular cluster can comprise up to a million stars.

Another difference is that open star clusters are found only in the disk of the Milky Way, whereas globular clusters are generously distributed throughout the halo. Since members of a star cluster were formed from the same gas cloud, they all have the same age and the same chemical composition, and are all equally far away from us.

Globular clusters are among the oldest objects in the Universe. Plate 6.G depicts M15, which is about 12 billion years old. What can these ancient objects teach us about the evolutionary history of the Milky Way? Decades ago, it was already possible to deduce the coarse structure of the Galaxy from the locations and age of open and globular clusters in our Milky Way. Put simply, one can imagine the Milky Way as having formed from a giant gas cloud in which huge areas slowly collapsed to form the central region. To remain stable, the gas began to coalesce to form a giant disk that rotates around the Galactic center. Ancient globular clusters indicate that this process likely took place a very long time ago. Since that time, countless stars have formed in open clusters located inside the disk in the gas-rich spiral arms that had emerged in the process.

Since open star clusters are relatively young, they are presumably still close to the place where they originally formed, namely the spiral arms. This suggests the spiral arms are regions with very generous reservoirs of gas and dust where new stars are frequently born. When the first distances to open clusters were determined, it became clear that they were residing along three kinds of rows weaved around the Milky Way's center. These curves were nothing else but segments of the spiral arms with the star clusters in it. This is how the spiral structure of the Milky

Way was initially discovered and mapped out—just by following the gas from which the open clusters had formed. Additional information about the sizes and distances between individual spiral arms could also be obtained from those cluster distances as well as the total size of the Milky Way. Today the detailed cartography of the spiral structure of the disk of the Galaxy is conducted with radio astronomy measurements that follow the hydrogen gas.

Star clusters play a major role in astrophysics. Since their members are all of the same age, are identically composed, and are equally far away from us, they are extremely well suited for studying a whole range of stellar characteristics, just as one would study people. Individuals differ a great deal, especially in their appearance. Schoolchildren in a class offer the possibility to witness the range of human characteristics in a sample of same-age individuals. Different school classes then illustrate the effect of human aging. In this sense, we can regard star clusters as the school classes of the Milky Way—they show us the full range of stellar traits and aging effects, especially when we compare different star clusters to one another.

Each star takes a particular position in the Hertzsprung-Russell diagram in accordance with its evolutionary stage. The details are described in chapter 4. Such a diagram can also be drawn for an entire star cluster, using a magnitude (a luminosity indicator) and a color (a surface temperature indicator) for each star. It is then called a color-magnitude diagram but is used in the exact same way as a Hertzsprung-Russell diagram. With some prior knowledge about stellar evolution you can determine the current evolutionary status of a star cluster from its color-magnitude diagram because all cluster stars arrange themselves in the familiar sequences and regions. A star is located higher up on the main sequence if it is more massive and luminous, otherwise it sits further down. The lifetime of a star depends on its mass: more massive objects have shorter life spans than the stars of lower mass.

The color-magnitude diagram of a very young star cluster, such as the Pleiades, shows a very prominent main sequence. These young stars, 115 million years of age, have not yet undergone any substantial evolution, so none of them are found in the area of the red giant branch. This behavior can be seen in Figure 6.4 (top).

A somewhat older star cluster, such as Praesepe, aged 730 million years, already exhibits some signs of stellar evolution in its color-

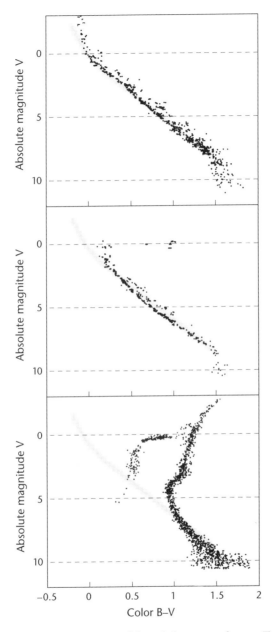

Figure 6.4. Color-magnitude diagrams of three different star clusters. *Top*: The Pleiades open star cluster is 115 million years old. *Middle*: The 730-million-year-old open star cluster Praesepe. *Bottom*: At 12 billion years of age, the globular cluster M15 is one of the oldest objects in the Universe. (*Source*: Peter Palm; data from J. C. Mermilliod, *Astronomy and Astrophysics* 53 (1976): 289–293 and J. C. Mermilliod, *Astronomy and Astrophysics* 97 (1981): 235–244 (for Pleiades and Praesepe); and Durrell et al., *Astronomical Journal* 105 (1993): 1420–1440, for M15)

magnitude diagram. In Figure 6.4 (middle), the upper part of the main sequence is not entirely populated anymore. Instead the more massive stars with their shorter life spans have already moved to the region of the red giants. Everything else in the lower section of the main sequence remains unchanged because the lower mass stars have not yet left the main sequence.

An old star cluster, such as the globular cluster M15, at 12 billion years, has only the lower part of the main sequence left. As can be seen in Figure 6.4 (bottom), all the stars, apart from those lowest in mass, have already evolved into red giants, resulting in a prominent red giant branch. Furthermore, some stars will surely have already exploded as supernovae and are therefore no longer present in the color-magnitude diagram.

The distance of a younger open cluster is easiest to determine when compared against a cluster of already known distance. By superimposing their prominent main sequences by shifting them vertically against each other, a difference in magnitude can be obtained between the reference cluster and the cluster under investigation. This magnitude difference is a measure of the difference in distance between the two clusters.

Distance determinations of old globular clusters are performed using RR Lyrae stars. Like the Cepheids, these stars periodically vary their brightness and are thus often referred to as pulsating cluster variables. In a color-magnitude diagram, they occupy a characteristic spot on the horizontal branch. The existence of these stars in the color-magnitude diagram indicates an already advanced age, since the horizontal branch is only reached in late stages of stellar evolution. They are particularly helpful in distance determinations because their actual luminosity can be obtained with the period-luminosity relation. At the same time, the observed magnitude of RR Lyrae stars can be read off the color-magnitude diagram on the horizontal branch (see Figure 6.4, bottom). The distance can be derived from the difference between the stars' observed magnitudes and luminosities.

For age determinations of open star clusters, the color value of the turn-off point from the main sequence in the color-magnitude diagram is needed. As described above, the color of the turn-off point in the color-magnitude diagram changes with the evolutionary state, and hence the age, of the whole star-cluster population. It becomes redder with increasing age because more and more of the more massive stars have already evolve away from the main sequence and in the direction of the red giant branch (see Figure 6.4). Thus, the turn-off point moves toward the right.

In general, cluster ages can be obtained using isochrones. These are the theoretical evolutionary tracks of stars in the Hertzsprung-Russell diagram or a color-magnitude diagram that have mainly the same age but different masses (see Figure 4.8). Isochrones are based on detailed computations of stellar evolution, that is, they track the temporal evolution of stars and how their observed properties such as luminosity and surface temperature change with time. A star cluster is thus nothing else but an "observed" isochrone, as all its stars share the same age but do not all have the same mass. However, the position and shape of the cluster stars in the color-magnitude diagram also depend on the chemical composition of the stars. A comparison of the positions of the cluster stars with isochrones of different ages, particularly in the turn-off point region, can lead to an accurate age determination only if the metallicity of the isochrones matches that of the cluster.

To determine the age of globular clusters, various sets of isochrones are needed with different metallicities and age levels. Globular clusters exhibit diverse metallicities and the right one has to be selected. Again, the relevant position of the main sequence turn-off in the color-magnitude diagram depends both on age and on the star cluster's metallicity. The positions of main sequence, red giant branch, and horizontal branch in the color-magnitude diagram of a globular cluster have to be shifted against those of the isochrones of relevant metallicity until one isochrone matches. This "correct" isochrone then immediately yields the age of the globular cluster.

Star clusters continue to be favorite objects of study—for ever more observations as well as theoretical studies, for instance, on the formation of globular clusters. Many newly discovered details constantly raise new issues. Improvements in data quality in the past 10 years are revealing small but important distinctions: In several star clusters, not all stars have precisely the same metallicity as is generally assumed of a cluster. This could be due to small inhomogeneities in the gas cloud from which the cluster stars emerged. Furthermore, other clusters contain up to five different subgroups of stars of different ages and metallicities.

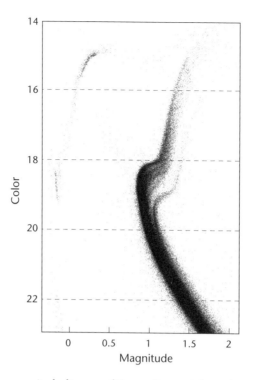

Figure 6.5. Color-magnitude diagram of Omega Centauri. Several populations exist in this unusual star cluster, as can be seen by the three main sequences. (*Source*: Peter Palm; data from Bellini et al., *Astronomical Journal* 140 (2010): 631. With the kind permission of Dr. Andrea Bellini)

As a consequence, some of the most massive globular clusters exhibit many separate main sequences in their color-magnitude diagrams, for instance, Omega Centauri in Figure 6.5. To take up the analogy of a school class again, it corresponds to quite a few failing students. If there are more failing students than regular students in the class, however, it becomes difficult for the observer to identify the actual grade of the class from just the students. Although many years have passed since their discovery, the underlying physical reasons for these different subgroups are still awaiting full explanation.

Finally, element abundances indicate that globular clusters have a chemical pattern entirely different from typical halo stars. These differences can be understood only as the outcome of some sort of "mud-slinging contest" inside the cluster: a kind of special chemical evolution specific to each cluster in which the early star generations produced metals that never left the cluster. These metals were then incorporated into the subsequent generations, producing unusual abundance patterns from then on. Stellar winds, composed of recently synthesized metals, could have further contributed toward the contamination of a cluster's gas. Later stellar generations would also have formed from chemically altered gas.

Globular clusters are thus somewhat like small countries with their own special laws. They are extremely complicated, and astronomers have a far from complete understanding of them. All in all, many different models of globular-cluster evolution have been developed, but no single idea seems to be able to explain all observations. Finally, the cosmological role played by globular clusters is also still unclear because it remains to be seen from which kind of gas clouds and under what conditions these immense objects could have emerged.

6.5 Naming Stars

I am often asked whether the stars that my colleagues and I find already have names or whether we can name them ourselves. The short answer is that all objects do actually already have names, such as their coordinates or some other combination of numbers or designation. The enquirer is usually visibly disappointed with this response. There is something to it, being able to assign attractive names to stars rather than some sober enumeration. The naming of celestial objects has not always been so boring, though.

Many of the brighter objects visible to the unaided eye have Greek, Latin, or Arabic names. The ancient Romans and Greeks knew about many stars and had names for them. For example, Ptolemy listed 1,022 stars by name in his catalog as early as the 2nd century BC. They liked to avail themselves of symbolic and visual analogies for the constellations, or drew connections to their mythologies by borrowing the name of a god. Many of these names are still in use now, even by scientists. Examples are the main stars in constellations, such as Aldebaran, the red giant star that is supposed to represent the eye of Taurus. Polaris, the North Star, which guides us northward, is another example.

Other brighter stars, galaxies, and nebulae also have names by virtue of having been known and loved for a long time.

Around 1770, the French astronomer Charles Messier compiled a long list of bright celestial objects that appeared as nebulous spots in his telescope. He cataloged about 100 star clusters, nebulae, and galaxies. They were simply numbered sequentially in the order of their discovery, starting with the Crab Nebula as M1, and so on. The M stands for Messier Catalogue. Especially well-known entries include the Andromeda galaxy M31, the Orion nebula M42, the Pleiades M45, and the Whirlpool galaxy M51. These designations are used by astronomers to this day. In 1888, the Irish astronomer John Louis Emil Dreyer published the New General Catalogue (NGC), which was substantially more comprehensive. It contains almost 8,000 entries and remains the standard catalog for non-stellar objects today. The enumeration of these objects begins at NGC 0001.

All subsequent observational efforts developed their own naming schemes to label their objects. Hence, quite a few other catalogs exist in which stars are recorded. Accordingly, all of my metal-poor Hamburg/ESO stars have a name beginning with the abbreviation HE, followed by a shortened version of their coordinates. Other surveys have named all their stars by the pertinent photographic plate in combination with a running number, for example. Obviously, there are countless ways to attach names or designations to stars.

Today, the International Astronomical Union officially takes care of the designation of cosmic objects and offers recommendations on best practices. Some years ago, for instance, it was proposed that all objects be named by their complete coordinates fully written out. Many of the huge sky surveys followed that recommendation because they often deal with millions of objects that all have to be uniquely labeled. The disadvantage is that these coordinates lead to extremely long and ugly numerical combinations. When working with individual stars it very quickly becomes unpractical.

One example is CD −38° 245, our very first representative of the most metal-poor star population, identified as such in 1984. Its discovery will be presented in section 10.1. This name was given to it nearly 100 years earlier, when it was catalogued for the very first time. "CD" stands for the survey name ("Cordoba Durchmusterung"), −38° for a particular slice in right ascension, and 245 is the running number of stars observed in the slice. With its coordinates fully spelled out, it would be called J00463619-3739335, certainly not a convenient name. Fortunately, it is possible to retain and use existing conventional names, such as in this case CD −38° 245. J00463619-3739335 is not the only other name for CD −38° 245, though. Since this star is relatively bright, it has been repeatedly observed and catalogued in many sky surveys, mainly photometric ones, over the years. Eleven other designations can be found at present, even if some of them are hardly distinguishable:

CD −38° 245 (Cordoba Durchmusterung survey, 1892)

SB 319 (Slettebak & Brundage list, 1971)

CS 22188-0048 (HK Survey of metal-poor stars, 1985)

GSC 07532-00548 (Guide Star Catalog for bright stars, 1990)

HE 0044–3755 (Hamburg/ESO Survey, 1991)

HIP 3635 (Hipparcos Satellite Catalogue, 1997)

DENIS-P J004636.1–373933 (Deep Near-Infrared Survey, provisory designation, 1997)

uvby98-003800245 (Photoelectric Photometry Catalogue, 1998)

2MASS J00463619–3739335 (Two Micron All-Sky Survey near-infrared survey 2003)

USNO-B1.00523-00009596 (US Naval Observatory survey 2003)

RAVE J004636.2–373933 (Radial Velocity Experiment survey 2008)

As is the case for CD −38° 245, it is common that the designation appearing in the first catalog is retained. Moreover, the brighter a star is, the more often the star has been observed. This is helpful because each new survey usually provides new supplementary information about a star. The 2MASS survey, for example, makes available photometry in the near-infrared range. The US Naval Observatory survey, in turn, determined the proper motions of many stars, which are important ingredients for their distance determinations.

HE 1327–2327 is another example in this regard. Until 2014, it was the most iron deficient star, with 1/250,000th of solar iron abundance. Its discovery and usefulness to astronomy is described in chapters 9 and 10. It is about 2 magnitudes fainter than CD −38° 245 and, accordingly, has just two other designations. HE 0107-5240, the first star with less than 1/100,000th of solar iron abundance, is substantially fainter still, and hence, is included in only one other catalog.

Today, a central astronomical database is used to compile all this information.[1] By entering an object's coordinates or name, a researcher can quickly find out what is already known about that object from any major survey and which scientific articles discuss the star. This database is extremely useful for determining whether an object has been spectroscopically observed yet, for instance, or whether it is listed in a particular catalog.

Despite all these resources, rediscoveries of already known objects do inadvertently happen from time to time, which is not only frustrating but can also cost a lot of telescope time. Over the course of my PhD thesis research, CD –38° 245 and a few other very metal-poor stars also appeared in my sample, but of course under their HE name. My delight at having identified an interesting star was, in fact, premature—I was soon disappointed when it became evident that it was "merely" a rediscovery of the already known star. However, there is also a positive side to this: I demonstrated that our new search method was capable of detecting metal-poor stars. In that respect, my rediscovery of CD –38° 245 was very satisfying. After all, it is one of the most metal-poor stars and hence a very good example of the objects I sought in the first place.

The more complicated a star's name is, the more it is shortened or simplified, at least in common usage, such as in conversations or conference presentations. This applies to "coordinate" names just as much as to all the others. Particularly when stars in larger samples are considered, my colleagues and I cannot always remember all those "phone-number" designations. For the sake of simplicity, we thus often refer to the first author of a publication and then name the star after that author. As an example, HE 1327–2326 has often simply been referred to as HE 1327 and occasionally even as the "Frebel star."

---

[1] SIMBAD, http://cdsweb.u-strasbg.fr.

CHAPTER 7
✦ ✦ ✦ ✦ ✦

TALES TOLD BY LIGHT

In priniciple, astronomy is a simple science. All that can be measured is the radiation that is emitted by various objects as well as a few particles such as neutrinos coming from outer space. Additional information can also be obtained from the study of meteorites and solar system material such as Moon rocks. Overall, this makes exploring the cosmos a challenging endeavor, and astronomers need to use sophisticated methods and high-tech facilities to squeeze every last bit of information out of this radiation. Before we are able decipher the chemical composition of a star, we first need to consider what data can be obtained of an object and how this information is processed. Radiation in the form of visible light is of principal interest when working with metal-poor stars. In the following, the discussion is focused mostly on this type radiation.

Astronomers can measure the number, direction, and energy of photons coming to us from an object. Consequently, the astronomer's most important tool is the telescope, with various instruments to collect light even from the faintest objects. Next comes the computer, which is utilized to record and process the observed data. Finally, theoretical research, done analytically or through simulations, complements any knowledge gained from the observed data and also helps with interpreting the findings.

In the end, it is astonishing how many details about a celestial object can be uncovered in this way. An initial glimpse at the night sky does not, by a long shot, reveal the properties of a cosmic object. By speaking the language of physics and allowing so-called spectroscopy to assist us are we ready to translate what light tells us about stars and other objects in the cosmos.

7.1 A Little Lexicon of Light

What we call light is nothing else but electromagnetic radiation. When its wavelengths fall within a particular range, it becomes visible to the human eye. It becomes "light." Each wavelength is perceived by the eye as a different color. Visible light ranges from violet-blue at 390 nanometers, through green at 510 nm and orange at 600 nm, up to dark red at 740 nm. The range of visible light is schematically represented in Plate 7.A. The colors of the rainbow present the full range of wavelengths we can see. When light of all these colors appears simultaneously, it is perceived as white. This is nothing else but radiation composed of all those wavelengths.

Electromagnetic radiation occurs in other wavelength ranges as well. All of us are familiar with X-rays, with which bones in the body can be made visible, microwave radiation used in the kitchen, radio waves, and also infrared radiation, which we can feel for example as heat coming from a flame. Ultraviolet rays cannot be seen, but the consequences of such irradiation are painfully noticeable in the form of sunburn. Figure 7.1 shows the full range of wavelengths with the different kinds of rays.

The wavelength determines the behavior of the electromagnetic radiation, which we will simply call "light" from now on, irrespective of the wavelength involved. To start with, light is wavelike and particlelike at the same time. As a result, photons—the particles that "carry" the light—have a specific energy, depending on their wavelength. This energy can be determined by measuring either the wavelength or the frequency of light. More energetic light has shorter wavelengths and thus a higher frequency. Correspondingly, longer wavelength, lower energy light has lower frequencies. In general, light can be described by its wavelength, intensity, direction, and the velocity of its propagation. The speed of light, about 300,000 km/s in the vacuum, is a well-known fundamental constant because nothing moves faster. Almost all cosmic objects emit electromagnetic radiation over a broad range of wavelengths or energies, including visible light. Various wavelength ranges—even ones far beyond the visible range—can be observed with different types of telescopes.

Transparency of the terrestrial atmosphere

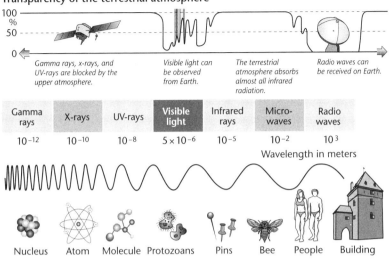

Figure 7.1. The electromagnetic spectrum from the gamma-ray range to visible light to long radio waves. The degree of transparency of the Earth's atmosphere to different wavelength ranges is illustrated above. This is why space telescopes are necessary for many observations. (*Source*: Peter Palm)

In the following, the term "body" refers to any object. It appears that everything around us "shines" in some way or another. The temperature of a body determines what kind of electromagnetic radiation is predominantly emitted. A cooler body radiates mainly but not exclusively at longer wavelengths than a hotter body. As the temperature rises, the maximum of the distribution of emitted light then shifts to shorter wavelengths.

For example, at 98.6°F, a human being is a relatively cold body. It radiates in the long-wavelength infrared range at 10,000 nm, which corresponds to 10 micrometers. Night vision devices are designed to operate exactly within this wavelength regime. Fireworks also radiate in this range. During an observing run on the summit of Mauna Kea on Hawaii on the Fourth of July, Independence Day in the United States, my colleagues and I were not just observing our target stars with a telescope but also the firework display on the beach in Hilo from an altitude of 4,200 m using night vision goggles. Fire emits mainly in the infrared region and only relatively little in the visible range. Conventional

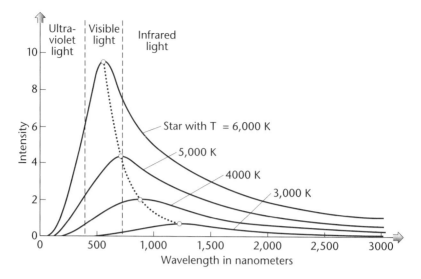

Figure 7.2. The spectral energy distribution of stars with different surface temperatures. At lower temperatures, the maximum of each distribution shifts to longer wavelengths. The Sun with 5,800 K has its maximum intensity roughly in the middle of the visible wavelength range. (*Source*: Peter Palm)

incandescent lightbulbs likewise shine mostly in the infrared and emit only about 10% of their energy as visible light. This explains why they get hot to the touch and why energy-saving lamps, which emit far less infrared light, are so much more efficient.

A star radiates in just the same way as any other body—with a different intensity at each wavelength and according to a characteristic energy distribution. The maximum of this energy distribution is a measure of the surface temperature of the star. This is shown in Figure 7.2. As an example, the Sun, having a temperature of about 5,800 K, radiates most of its energy in the middle of the visible wavelength range at about 500 nm, that is, in the blue-green region. We can thus perceive only 40% of sunlight with our eyes. The remaining 60% is radiated away in other wavelengths, as can be seen in the figure.

Cool red giant stars radiate most of their energy away in the red and infrared wavelength ranges. If the Sun were a red giant and were cooler, it would also be mainly emitting red rather than orange light. In this context, it should briefly be noted that the Sun does not turn into a cool red giant every evening and morning at sunset and sunrise just because it appears red to us at those times. In these instances, dust in the Earth's atmosphere causes the blue light to scatter out of our line of sight, leaving behind only the yellow-red portion to reach our eyes. Hence, this effect has nothing to do with the wavelength range of the light being emitted by the Sun itself.

Using special filters, particular wavelength ranges can be isolated when observing a star. The resulting brightness is then given with respect to the particular filter. As an example, a red giant appears much fainter in the ultraviolet range than in the red range because the star hardly radiates at those shorter wavelengths. Figure 7.2 illustrates the so-called spectral energy distribution of a star. The difference between two brightnesses, each of them averaged over the wavelength range covered by the filter, is called "color." These colors reveal how steep the energy distribution curve is and what the surface temperature of the star is. Together with a known surface gravity, the temperature then reveals information about the star's evolutionary stage and its characteristics.

In astronomy, the brightness of a star (and all other celestial objects) is expressed in so-called magnitudes. According to the historical definition, the brightest stars are of first magnitude and the faintest stars, still barely visible to the naked eye, of the sixth. Strangely enough, this means that fainter stars have greater apparent magnitudes. Pursuant to this historical designation, the modern definition states that the brightnesses of two stars that differ from each other by a factor of 100, differ by 5 magnitudes. This way, the full Moon and the Sun have negative apparent magnitudes, namely, -13^m and -27^m, because they are considerably brighter than the stars that originally formed the basis of this system. Sirius, the brightest star in the night sky, has -1^m, Jupiter has -3^m, and Venus, our evening or morning star, has -5^m. Venus thus appears to be 40 times brighter than Sirius. The Sun shines about 15 billion times brighter than Sirius.

There are many different methods to comprehensively analyze starlight, but they all have one thing in common: they all require photons to be collected with a telescope. Astronomical observations are divided into two types: imaging and spectroscopy. Imaging the sky serves to measure stellar magnitudes, positions, and spatial shapes. Repeat imaging over longer time frames also allows to track the motion of objects with respect to each other on the sky. Spectroscopy allows broad wave-length ranges of light to be decomposed into what is referred to as a spectrum. This kind of analysis permits determinations of the physical properties of an object, such as surface temperature and gravity, chemical composition, and its velocity in space.

7.2 Spectroscopy—Deciphering Starlight

A rainbow across the sky is an impressive natural phenomenon and its array of colors is always fascinating. During such an event, the different wavelengths of white sunlight are refracted and reflected differently inside the many tiny water droplets in the atmosphere. As a result, we see how the light is divided into a rainbow right before us. A simple prism works the same way. As illustrated schematically in Plate 7.B, the wavelengths of light are refracted by varying degrees at the glass surface but are not reflected. Shorter wavelength violet light is refracted the strongest, whereas longer wavelength red light is less so. White light is separated into its spectral colors as a result.

Astronomers have long utilized this technique of decomposing light and recording its spectrum. As early as 1800, Fraunhofer was able to recognize important details about the nature of the Sun this way. Spectroscopy was established soon afterward as a method of scientific investigation. All kinds of spectrometers and spectrographs were developed that were soon found in better-equipped laboratories. In astronomy, spectroscopy suddenly enabled detailed studies of the physics and chemistry of distant celestial bodies for the first time. Spectroscopy turned the distant and hard to reach stars into specimens that could be analyzed in the laboratory. Modern astrophysics would be inconceivable without the use of spectrographs.

Modern spectral analysis is still based on the same principles that Fraunhofer introduced. A prominent example is the pair of Fraunhofer lines of calcium, also labeled H and K. They are two of the strongest, most easily identifiable lines in most stellar spectra and are thus helpful in orientating oneself in a spectrum full of spectral lines.

What makes these spectral lines so fundamentally important? Section 2.2 also dealt with spectroscopy, but primarily with respect to its historical aspects. Let us now take another short detour into atomic physics to consider what exactly happens to starlight along its path to the spectrograph. For better readability, the following discussion will contain the occasional repetition of previously described topics.

If you analyze the light of a glowing, solid or liquid, body spectroscopically, such as the tungsten filament of a lightbulb or molten iron, you obtain what is called a continuous spectrum that displays all the colors of the rainbow. A sample is shown in Plate 7.C. Looked at more closely, a stellar spectrum looks just a little different. As Fraunhofer had already observed, there are dark lines in these spectra. No wonder, since a star with its extended gaseous atmosphere is more structured and complicated than a simple glowing body. The star's light comes from its hot center, where nuclear reactions are occurring. Along its path to us the light therefore passes through a thick layer of cooler gas in the outer stellar atmosphere. But this does not happen without some losses.

Let us recall how an atom is structured. The positively charged nucleus is surrounded by a swarm of negatively charged electrons. In the case of a neutral atom, there are as many electrons as protons in the nucleus, so overall the atom has a net zero charge. The electrons orbit at different levels corresponding to their energies. Instead of speaking of orbits that are too reminiscent of the planetary orbits in Bohr's model of the atom, we should rather refer to orbits as energy levels. Low-energy electrons mostly stay close to the nucleus, while electrons of higher energies move around farther away. Usually, all electrons are at the lowest possible energy levels. The lowest energy level is the so-called ground state that is most energy efficient for the atom. When light from the hot stellar center moves through the cooler stellar atmosphere, electrons in the atoms in the atmosphere begin to jump onto higher energy levels. The photons from the stellar interior have all kinds of energies. When one of them collides with an atom with the right energy for an electron to jump between two energy levels, the electron takes advantage of the situation and indeed jumps to the next higher level. The photon disappears in the process. Each electron jump corresponds to a very specific energy, and hence a very specific light wavelength that prompted the jump.

Jumps between energy levels lead to photons of very specific energies—or wavelengths—being used up while the others are not affected at all. After passing through the stellar atmosphere, the starlight thus suddenly "lacks" certain wavelengths. Precisely this effect is observable in stellar spectra: the dark Fraunhofer lines are nothing else but the missing wavelengths that the light had to pay as a "toll" to leave the star. Astronomers call these lines absorption lines. How an absorption line spectrum arises is schematically depicted in Plate 7.C.

This process also works in reverse. So-called emission spectra arise from a uniformly hot gas. There, all electrons are at high energy levels. From time to time, electrons jump down to a lower level whereby they release the corresponding energy in the form of a photon. As a result, the light emitted by the gas acquires additional energy at the relevant wavelengths, which is observed as emission. Spiral galaxies, irregular galaxies, and various other types of galaxies, as well as planetary nebulae exhibit many emission lines since they resemble a giant cloud of gas. The way an emission line spectrum develops is represented in the middle diagram of Plate 7.C.

Although Gustav Kirchhoff did not know anything about the details of atomic structure and energy level transitions, he could already distinguish among three main types of spectra on the basis of his experiments around 1860. He could predict whether to anticipate a continuous spectrum, an emission spectrum, or an absorption spectrum following the observations of different light sources. Bohr's atomic model was eventually able to physically explain the behavior of spectral lines. It was yet another step on the road to quantum mechanics. Kirchhoff's three rules for the formation of spectra are just as relevant today as they were 150 years ago. They are used particularly in analytic chemistry and of course astronomy. Generally, all stars produce absorption spectra unless they have an active surface layer or else are enveloped in hot gas. This can happen in a binary-star system, for instance. Then, emission lines attributable to the hot gas are visible in addition to the absorption lines.

Astronomical observations can examine only the surface layers of a star. This also applies to spectroscopy. The stellar interior remains hidden, exactly as a person's internal organs are invisible behind her or his skin. Fortunately, there are other means of gathering important information about what is happening inside a star, although we will not discuss them further.

With spectroscopy, specific properties of the stellar atmospheric gas can be measured. Each chemical element exhibits a characteristic absorption line pattern in the spectrum. The absorption lines at specific wavelengths in a spectrum therefore reveal which elements are present in the atmosphere. Hydrogen lines of the so-called Balmer series are the most important and often also the strongest lines in a stellar spectrum. The strongest of these lines, called Hα (H stands for hydrogen, α for the strongest of these hydrogen lines), is located at a wavelength of 656 nm in the red spectral range. The entire series of lines forms when electrons from the second level of the hydrogen atom jump to other levels further out (see also Figure 2.3). The transition from the second to the third level requires a photon with a wavelength of 656 nm. Many pretty nebulae are of red color due to hydrogen, and in particular the Hα emission line. The Fraunhofer calcium lines H and K are extremely prominent as well. They are at 397 and 393 nm in the violet spectral range. The calcium H line almost completely overlaps with one of the Balmer lines, Hε. In a star's atmosphere elements can occur in neutral as well as ionized form. Calcium H and K are attributable to ionized calcium. The strongest line of neutral calcium, however, is at 423 nm. In addition, three characteristic magnesium lines, around 528 nm in the green, and two prominent sodium lines, around 589 nm in the orange range, can be encountered in almost every stellar spectrum. Depending on the star type, there are countless lines of various other metals. Identifying lines correctly in a spectrum relies on detailed comparison data of the various elements. Indeed, based on extensive studies in the laboratory, information on many thousands of spectral lines and details about the atomic physics of the associated transitions fill voluminous catalogs and comprehensive data bases.

Overall, the physical conditions in the stellar atmosphere determine whether a line appears in the spectrum and how strong it is going to be. For example, line strengths depend strongly on the temperature. Hydrogen lines are most prominent in stars with surface temperatures around 10,000 K. However, in cooler stars such as the Sun, at about 5,800 K and located on the main sequence, the Balmer series is much weaker in the spectrum although still significant. The opposite applies to metal lines. They are so weak at higher surface temperatures that they can hardly be detected. Interestingly, helium lines begin to appear in the spectra of stars with surface temperatures of at least 10,000 K. In cooler stars, the temperature is not high enough for the excitation of helium atoms.

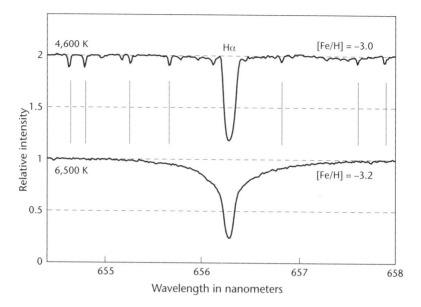

Figure 7.3. Normalized spectra of two stars with different temperatures but the same metallicity. The line strengths differ despite the same metal abundances. The upper spectrum has been shifted for easy comparison. (*Source*: Peter Palm; spectra from Anna Frebel's private collection)

The metal lines can be rather strong, though. This behavior is shown in Figure 7.3. Cool red giants have narrower, weaker hydrogen lines, but in turn much stronger metal lines. Stars that are even cooler, for instance, with just 2,000 K, have bands of countless overlapping titanium-oxide lines in the spectrum. At those low surface temperatures, the formation of molecules is possible in the stellar atmosphere. These behaviors are independent of the elemental abundances but result from temperature and gas pressure effects.

As a result, every spectrum is a star's "fingerprint" from which it is possible to deduce its nature and characteristics. The first so-called spectral classes were introduced before 1900 to bring order to the multitude of spectra. Soon afterward, a more refined OBAFGKM sequence with additional subgroups was proposed. At the same time, thousands of stars were being classified by Annie Jump Cannon (see also section 2.3). It turned out that spectral classes based on different line strengths essentially represented a surface temperature sequence. The spectral classes

Table 7.1. Relation between Spectral Type and Temperature

Spectral type	Color	Temperature (in K)
O	Blue	30,000–50,000
B	Blue, bluish white	10,000–3,000
A	White	7,500–10,000
F	Yellowish white	6,000–7,000
G	Yellow	4,500–6,000
K	Reddish orange	3,500–4,500
M	Red	2,000–3,000

still in use today and their corresponding stellar temperatures are listed in Table 7.1. O-type stars have surface temperatures of 40,000 K. M-type stars at less than 3,000 K are cool by contrast.

Modern spectral analyses have demonstrated that most stars contain a large number of elements in their atmospheres and that their metallicities very much resemble that of the Sun. Nevertheless, metal-poor

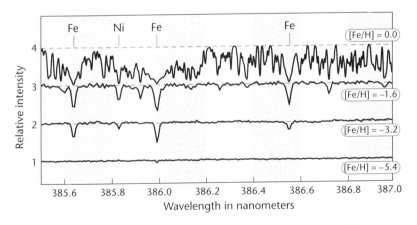

Figure 7.4. Spectra of stars with similar temperatures but different metallicities. *Top*: The metal-rich Sun with [Fe/H] = 0. *Middle*: Two stars with lower metallicities. *Bottom*: The star with the second lowest iron abundance, [Fe/H] = –5.4. The most iron-poor star does not show any Fe lines in the spectrum anymore. The meaning of "[Fe/H]" is described in section 7.3. All the spectra are normalized and shifted for easy comparison. (*Source*: Peter Palm; reproduction of spectra from Frebel, *Astronomische Nachrichten* 331 (2010): 474–488)

stars do also exist with metallicities much lower than the Sun's. What do the spectra of stars with different metallicities look like? The difference is immediately apparent when you look at two equally hot stars: one as rich in metals as the Sun, and 100 times more metal deficient. The metal-poor star has significantly weaker metal lines, as can be seen in Figure 7.4. As the concentration of elements in the atmosphere is lower for metal-poor stars, there are fewer atoms of those elements that would absorb light at the wavelengths characteristic of the transitions between the energy levels in the atom. Also, hotter stars of the same metallicity have weaker metal absorption lines than cooler stars because more gas is ionized. If we want to find metal-poor stars, we have to look for stars that display very weak metal lines particularly in the cooler stars.

7.3 Element Abundance Analyses of Stars

The goal of stellar archaeology is to determine the detailed element abundance patterns of the most metal-poor stars. The somewhat cumbersome term "element abundance analysis" describes the process that simply reveals what the outer layers of a star are made of. Only then can metal-poor stars be identified as such, ultimately providing information about the young Universe and the processes of nucleosynthesis during that era.

For a complete abundance analysis you first need a high-resolution spectrum of the star to be studied. High spectral resolution is of great important to resolve the many weak elemental absorption lines, that is, to make them cleanly measurable. If the resolution is too low, these spectral lines will be smeared out and become undetectable. But acquiring such a spectrum is easier said than done: most of the time the star is faint, and a large telescope and long exposure times are needed to obtain a spectrum (the signal) of sufficient quality, that is, one with a clear signal and little noise, such as noise generated by the camera. Figure 7.5 shows spectra of a star with different signal-to-noise ratios and the corresponding exposure times. It is immediately evident that longer exposure times lead to a better spectrum with a higher signal-to-noise ratio. Unfortunately, due to statistical reasons, the signal-to-noise ratio increases only with the square root of the observing time: the exposure

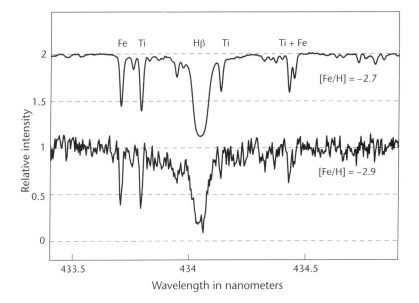

Figure 7.5. Normalized spectra of stars with different signal-to-noise ratios (S/N) in the region of the Hβ-line. *Top*: A spectrum with a high S/N of ~300. *Bottom*: A spectrum with a S/N of ~13. To obtain the high S/N, the fainter star would have to be observed 530 times longer or else be seven magnitudes brighter. The upper spectrum is shifted for easy comparison. (*Source*: Peter Palm; spectra from Anna Frebel's private collection)

has to be four times as long for the signal-to-noise ratio to become twice as good. This fact explains why spectra are often "noisy," particularly those from fainter stars, because better data would require prohibitively long observations.

How can the chemical composition of a star be determined from its spectrum? The first main task of the analysis is measuring the strength of the absorption lines in the observed spectrum. Mathematically, the strength of a spectral line is measured as a so-called equivalent width and describes the area covered by the line. The equivalent width then refers to the width of a rectangle with a height of the entire line and area that is equal to the area of the spectral line. Equivalent width does not change with spectral resolution, but at high resolution the line is deep but narrow and sharp whereas it is shallow and smoothed out at lower resolution.

Line strength is ultimately a measure for the number of atoms of a given element that is present in the stellar atmosphere, although other input is also required when determining an abundance. However, before the lines in a spectrum can be measured with a computer program, you first need to know which lines belong to which elements. Thanks to comprehensive catalogues of the physical properties of atoms, it is possible to look up the wavelengths at which every atom absorbs. In the end it thus relatively easy to identify the observed lines in a stellar spectrum. Once you have found lines of a particular chemical element in the spectrum, it is proof that this element does in fact exist in the stellar atmosphere. Later, these line measurements have to be transformed into actual abundances with the help of a stellar atmosphere model.

In practice, though, this whole procedure is actually done the reverse way. We know at which wavelengths important spectral lines occur. So we look for these particular lines in a spectrum for later measurement. For example, each new spectrum is immediately checked at the telescope for example for neutron-capture elements such as strontium, barium, and europium. Strontium has its strongest line at 4,077 Ångström, barium at 4,554 Ångström, and europium at 4,129 Ångström. The majority of metal-poor stars display only noise at the spot where the europium line ought to be. The presence of a strong europium line indicates that the star is most likely an r-process star or otherwise of interest. But even particularly strong or nonexistent strontium and barium lines may suggest a very interesting nucleosynthesis origin for the elements present in the star.

And what are Ångströms? According to the International System of Units convention in physics, wavelengths of light should be given in nanometers (nm). Astronomers, however, still follow tried-and-true historical tradition and denote the wavelengths of spectral lines in "Ångström" (Å for short). Ten Ångström correspond to one nanometer. Astronomers do not seem to like decimals very much. Using the Ångström unit is more practical and avoids very small numbers or too many digits after the decimal point, which would be unavoidable using nanometers. For example, the important uranium line (see section 5.3 and Figure 5.8) has a wavelength of 3,859.57 Å on the Ångström scale, which is 385.957 nm.

This difference in notation may seem rather trivial—nonetheless the Ångström convention is actually quite helpful. The weak uranium line is positioned closely to three other lines. These neighboring lines are an iron line at 3,859.21 Å and a neodynium line at 3,859.43 Å on the left-hand, shorter-wavelength side of the uranium line, as well as a strong iron line at 3,859.91 Å on the other side. Their wavelengths merely differ by a few digits after the decimal point because they are so close together and even partly overlap. In nanometers, these wavelengths would differ only in the second digit after the decimal point. Even the third digit after the decimal point would then still be of importance when describing the precise position of a line.

Another advantage of using the Ångström unit becomes evident when measuring absorption lines strengths. The weakest spectral absorption lines are typically only a few milli-Ångström (0.001 nm) wide, whereas the stronger ones can be 200 milli-Ångström (0.02 nm) or more. Examples of different absorption lines can be seen in Figure 7.5. A few lines, such as the calcium K line, are much stronger, with line strengths of 1 Å or more. However, for reasons involving the physics of stellar atmospheres, these strong lines are less useful for any detailed abundance determinations based on high-resolution spectra.

Compared to other elements, iron has the advantage of possessing a plentiful absorption line spectrum. This means that numerous iron lines are found throughout the entire visual spectral range. The more metal-rich stars have thousands of detectable lines, whereas stars more metal-poor have no more than some hundreds of such lines left. In the case of hotter metal-poor stars, only a handful of iron lines are visible in the spectrum. For various reasons, not all these lines can be measured. Surprisingly enough, this is particularly true in metal-rich stars because the many strong lines all blend together leaving only a few clean lines for measurement. Nevertheless, iron is the element for which by far the most lines can be measured. As a result, the stellar metallicity can be precisely determined especially because one does not have to rely on a particular spectral range for measurement. Furthermore, having many individual measurements available leads to a smaller uncertainty in the final iron abundance.

This is enormously important because with a few exceptions the iron abundance of a star describes its total metal abundance—the abundance of all elements apart from hydrogen, helium and lithium. For this reason, iron serves as a reference element for all other chemical abundances. Astronomers love to calculate abundance ratios rather than absolute, individual abundances because abundance *ratios* have one big advantage: among other things, various systematic measurement uncertainties cancel out.

In general, absorption lines of 15 to 20 elements are identifiable and measurable in the spectra of metal-poor stars. This includes lines of CNO elements, α-elements, and elements in the iron-group (see also Table 3.2). The neutron-capture elements strontium and barium are found in most metal-poor stars as long as the surface temperature is not too hot. It would make the line strengths become too weak for detection. Then, abundance ratios are formed, such as C : Fe, O : Fe, Mg : Fe, and so on. In some cases ratios such as C : O and Sr : Br are also useful for testing certain nucleosynthesis predictions.

How are the values for element abundances obtained from line-strength measurements? Complex computer programs that simulate the stellar atmosphere are required. The outer layers of a star are modeled to calculate the physical processes that cause the absorption seen in the spectrum. Stellar atmosphere models can be characterized by the parameters surface temperature, surface gravity, and chemical composition. Based on those values, other quantities, such as the pressure, are calculated in different layers of an atmosphere as they can vary greatly between stars. In the atmosphere of a red giant star, for example, the surface gravity is low. Hence, the pressure is lower compared to that of a main-sequence star. The tricky bit is knowing which atmosphere model best matches the physical characteristics of that star. Finding the correct model is an iterative procedure and a major part of the analysis. For that, the stellar parameters surface temperature, surface gravity, and the star's metallicity need to be obtained. One starts with a guess for all parameters and then works toward the final values through a series of steps and tests. With some experience, inspecting the observed spectrum of the star to be analyzed tells you immediately whether you are dealing with a giant or a main-sequence star. Still, more detailed work is needed.

As previously indicated, knowing the stellar surface temperature is fundamental for the element-abundance determination. The line strengths in the star's spectrum depend not only on chemical composition but also on the surface temperature. If the wrong temperature is used in a model, the abundances will be incorrect. The same applies to the surface gravity, although to a lesser extent. The temperature can either be determined by using various stellar colors (e.g., B-V) in combination with temperature calibrations or by making use of the fact that the abundances of the individual iron lines must all be equal as a function of their excitation potential. Otherwise different lines would yield very different iron abundances, even though a star can of course have only one single iron abundance. A similar argument is applied when determining the surface gravity. Due to a connection between the gas pressure, surface gravity, and ionization of the atoms, the absorption line strength of singly ionized iron also depends on the surface gravity, while those of neutral iron lines do not. Nevertheless, the lines arising from neutral iron atoms must yield the same iron abundance as those from ionized iron atoms. The surface gravity is found this way.

In the end, the procedure is a bit like a jigsaw puzzle. After a few tests and trials you eventually reach the goal and have obtained the star's surface temperature, surface gravity, and finally its metallicity. With the correct parameters, the model atmosphere does, in the end, match those of the observed star. The individual abundances of all the elements whose lines have been measured in the spectrum can finally be calculated.

In practice, the procedure for determining element abundances is also performed in reverse order. Many lines in the spectrum overlap or form complicated absorption features. Determining an abundance from such an blending line cannot be done through the direct line-measurement technique. From the analysis of iron line strengths, it is already known which model atmosphere best describes a star. This fact can be utilized here: A synthetic spectrum that imitates the observed spectrum can be generated with that model atmosphere. Any already found element abundances are adjustable in the model to change the extent of the absorption in the to-be-synthesized spectrum with the goal of matching the synthetic spectrum to the observed one. This way, a complicated absorption line region in a star's spectrum can be "built." Figure 7.5 as well as Figure 5.8 depict such a spectral region, together with the best matching synthetic spectrum. The individual abundances of the participating elements can be determined this way, despite any blending with neighboring the lines.

In the end, all measured abundances describe the number of atoms of an element in the stellar atmosphere as it relates to the number of hydrogen atoms. A practical notation for the abundance ratio of two elements A and B has become the convention: $[A/B]$. This "bracket notation" expresses that the abundance ratio has been determined relative to the Sun. Mathematically, the definition of the bracket notation is as follows:

$$[A/B] = \log_{10}(N_A/N_B)_{star} - \log_{10}(N_A/N_B)_{Sun}.$$

At first glance this equation looks complicated. But it describes nothing more than a logarithmic ratio between the number of atoms N of element A, divided by the number of atoms of element B in the star's atmosphere compared to that of the Sun.

Let us look at a couple of examples that illustrate that this notation is very practical and simple to understand. The iron abundance is employed as an indicator of the metallicity of a star. For that, the ratio of iron to hydrogen is determined and then compared with the solar value. The corresponding bracket notation looks like this: [Fe/H]. A star whose iron-to-hydrogen ratio is the same as the Sun's is assigned a metallicity of [Fe/H] = 0. Metal-rich stars similar to the Sun accordingly have [Fe/H] ~ 0.

If a star contains twice as many iron atoms per hydrogen atom than the Sun, its metallicity is [Fe/H] = 0.3 (not 2 because the notation is logarithmic). If, on the contrary, the metallicity lies below 0, the star contains fewer metal atoms relative to the number of hydrogen atoms than the Sun and consequently counts as metal-poor. A star with [Fe/H] = −1 has 10 times fewer iron atoms relative to hydrogen than the Sun. Extremely metal-poor stars have [Fe/H] = −3 or less. Table 3.1 already indirectly contains this information, since −3 is the logarithmic value of 1/1,000th of the solar iron value. Stars with [Fe/H] = −5 have just 1/100,000th of the solar iron value; and so forth. Overall, astronomers have found [Fe/H] values ranging at least between −7.3 and +0.5 for stars in the Milky Way and various dwarf galaxies.

Whereas the [Fe/H] value describes the metallicity of a star, other ratios, such as [Mg/Fe], indicate how other elements behave in relation to iron and the Sun. If [Mg/Fe] = 0, the ratio of magnesium to iron is exactly the same as in the Sun. If [Mg/Fe] = +1, than compared to the Sun, the star has 10 times more magnesium than iron.

It thus follows from the definition of the bracket notation that the metallicity relates not only to the absolute ratio of a given atom in the stellar atmosphere but also to the corresponding value of the Sun. That is why astronomers are constantly using the Sun as a reference star. It is also clear that stellar abundances are directly related to solar abundances. If the latter change, so do the values for all stars. From time to time solar abundances are, in fact, redetermined. These changes do not reflect any real changes in the Sun. They rather show how difficult it is, even today, to measure the solar abundances, and stellar abundances more generally, with very high precision.

What happens next after the element abundances of a star have been determined? They permit a detailed reconstruction of the nucleosynthesis processes that operated in the previous stellar generations, thereby driving chemical evolution. Results on the nature of the very first stars in the Universe, acquired through analysis of the most iron-poor stars, are presented in section 9.1. Then, section 9.4 explains how the abundances of many different metal-poor stars with different metallicities are used to piece together the evolution of the lighter elements in the Milky Way. The origin and evolution of neutron-capture elements, that is, the heaviest elements, have been examined in chapter 5.

7.4 The Largest Telescopes in the World

We are looking for the oldest stars from the era shortly after the Big Bang, hoping that they will reveal insights into the evolution of the early Universe, the first generations of stars and the onset of chemical evolution. Yet, without large-scale observational programs, these rare ancient stars cannot be found in the vast cosmos. Telescopes on every continent are being employed in this search. Today, multinational projects are very common in astronomy, as scientists from many countries are working together in large teams. Large telescopes are also often operated by international consortia.

Light is always in short supply when it comes to astronomical research. For that reason telescopes should be as large as possible to capture as many photons from cosmic objects as possible. The largest telescopes for astronomical observations in the optical and also infrared wavelength ranges have primary mirrors measuring 6.5 m to 10 m in diameter. In astronomy, the term "optical" refers to the range of visible light. Which wavelength range is covered in a given observation then depends on the individual instruments. There are various instruments for particular kinds of optical and near-infrared observations.

The wavelength range observable from Earth extends from about 310 nm to a few micrometers. Radio waves with wavelengths of millimeters and longer are also observable. The terrestrial atmosphere prevents observations in the UV range at wavelengths shorter than 310 nm. Life on Earth benefits from it, though, since the otherwise harmful ultraviolet rays would reach Earth unhindered. In that case not even sunscreen lotion could protect our skin anymore. Observations even farther into the short-wavelength UV range are therefore carried out using space telescopes with special detectors. These include observations of X-rays and gamma radiation with their very short and energetic wavelengths.

For observations beyond the near-infrared range, optical-infrared telescopes should best be positioned on very tall mountains with extremely dry air, or better yet, in space. After all, the water vapor in the terrestrial atmosphere absorbs large amounts of infrared radiation. Radio telescopes are necessary for observations of radiation with even longer wavelengths. Instead of using reflecting mirrors, this type of telescope uses huge dish antennas with diameters of up to 100 m to receive cosmic radio radiation.

The technical effort necessary to manufacture an optical mirror is enormous. Few workshops and laboratories in the world can meet the extraordinary technical requirements to make mirrors many meters in diameter. Since the cost rises steeply as the diameter increases, some of the largest telescopes are not equipped with a single giant mirror but rather with a whole set of smaller mirrors that are fitted together to form a honeycomb shape. It is technically very difficult to produce and operate individual mirrors larger than about 8 m in diameter, so the designs for future generations of giant telescopes with main-mirror diameters of up to 40 m are all composed of multiple mirror segments.

Having a large mirror is not everything, though. It alone does not guarantee good observations. The sites for telescopes have to be very carefully selected. Climate conditions as well as environmental and political considerations are extremely important for the success of a project. In general, tall mountains and plateaus in dry regions with stable climate are favorable for all observations in the optical regime. Apart from the low humidity, the high altitude air is clear and clean. For astronomers it is especially important that the airflow at and over the telescope site be as smooth and free of turbulence as possible—otherwise, when looking through the telescope, the objects will appear less sharp. In addition, it is important that observatories lie far away from urban areas so that observations are not affected by light pollution. And finally, the country chosen for the site also plays a role. It should be politically stable because telescope projects are long-term commitments. They need substantial local and regional support, especially during the construction phase.

The two Magellan Telescopes (each with a mirror diameter of 6.5 m) at Las Campanas Observatory are operated by the American Carnegie Institution for Science. A two-hour drive away from the cities of La Serena and Coquimbo, they are located at an altitude of about 2,500 m in the Chilean Atacama Desert, some 600 km north of Santiago de Chile (they are shown in Plate 7.D). The four identical 8-m telescopes of the European Southern Observatory (ESO) are a two-hour flight north of Santiago, near Antofagasta on the Paranal summit in northern Chile.

The climatic conditions in this arid desert landscape are the best, from the astronomical point of view, with constant stable temperatures ranging from 50 to 60 °F, low humidity, and very little precipitation. The 4,200-m high peak of Mauna Kea on Hawaii's Big Island is another astronomer's paradise with ideal weather conditions. The Japanese Subaru Telescope (8 m) and the pair of American Keck Telescopes (each 10 m, but with segmented mirrors) both are situated there and can be seen in Plate 7.E.

There are also the two 8-m Gemini Telescopes, one of which is on Mauna Kea in Hawaii and the other near La Serena, Chile, where it is operated by the Cerro Tololo Inter-American Observatory. The Large Binocular Telescope possesses a segmented 8.4-m mirror and is located in the US state of Arizona. The Gran Telescopio Canarias with its 10.4-m segmented mirror was built on La Palma, Canary Islands. These telescopes are not, or perhaps are not yet, equipped with optical high-resolution spectrographs, and for that reason are not employable for the search for metal-poor stars.

Finally, there is the segmented 6.5-m MMT in Arizona and the Hobby-Eberly Telescope in the western part of Texas. It also has a segmented mirror with a total diameter of 9.2 m. Its less flexible construction restricts its maneuverability. As a result, objects can be observed only for relatively short periods, and the full surface of the mirror can often not be employed for observation. A twin of this telescope is located in South Africa (South African Large Telescope, SALT). In addition, there are countless smaller telescopes on all the continents, including Antarctica.

As already indicated, every telescope has different kinds of attachable instruments for astronomical observations of specific wavelength ranges. These instruments can generally be divided into two classes. First, there are so-called imagers that work like gigantic digital cameras and essentially photograph regions of the sky. Today, to view the picture being taken by a modern digital camera, one looks through not a viewfinder but a small screen. It is just the same with professional telescopes. Any new astronomical image is read out from the chip and immediately displayed on a computer screen. Using software, the observer is then able to immediately judge whether the exposure was long enough or if more data are needed. These images enable measurements of an object's brightness (photometry) and/or position (astrometry), and also spatial shape for example in the case of galaxies. There are different kinds of imager instruments that are used for specific observations, such as wide-angle exposures, detailed images, and exposures with particular filters covering broad or narrow wavelength bands to observe specific "colors." It all depends on what is needed to solve the particular scientific problem.

The other class of instruments is spectrographs, which split the light into a spectrum. Again, a raw spectrum is displayed on the computer screen after each exposure. Plate 7.F shows such a just-taken raw spectrum at the Magellan Telescope. Absorption lines are discernible in the original exposure as dark lines (many are the Fraunhofer lines). Such an absorption (or emission) line spectrum needs to be processed first with specifically designed computer programs to remove instrumental and detector signatures and subtract sky background light to reveal the actual measurement if the object. Then, a summed up crosscut of the spectrum image yields the final extracted spectrum ready for a detailed analysis. Examples of other raw spectra can also be found in Figure 2.1.

In the past two decades it has become possible to observe many hundreds of objects at the same time with "multi-object" spectrographs, provided they are close enough to each other on the sky. With specially designed slit plate or using so-called fiber positioners, multiple objects can be observed simultaneously. The light of each object either falls through its designated slit in the plate or is collected by optical fibers and then sent to the spectrograph. Example observations include members of a globular star cluster or stars in small dwarf galaxies. This new observing technique drastically reduces the overall observing time, especially when large numbers of objects are required. The disadvantage of this multi-object spectroscopy usually is that all the objects can be observed only in a limited wavelength range. This can become problematic when working with metal-poor stars because of the relatively low number of available spectral lines. As most of the metal-poor halo stars are too far apart on the sky, this technique is of interest mostly with respect to dwarf galaxy stars.

Astronomical images, regardless of whether a spectrum or a picture, taken with these instruments are recorded on a detector chip, which is nothing but a huge camera installed inside the instrument. This detector chip is a so-called CCD (charge-coupled device) and, in principle, is equivalent to the chip inside a typical digital camera. A CCD can be imagined as a kind of chessboard, with each of its squares equipped with an electron counter. Photons coming from a star, for instance, hit some fields of this CCD. There they release electrons by means of the photoelectric effect. The chessboard squares correspond to small picture elements, called pixels. Suitable computer software then reads off how many electrons have been released in each pixel and how many photons correspond to it. Consequently, astronomers basically observe photon counts. This sounds pretty unromantic if your grand aim is to study the Universe. But it is exactly this and only this information that we can gain from those distant cosmic objects. Observational astronomy as a whole is based upon this technology.

Modern digital photography for the common user is actually an example of how society benefited from progress made in astronomical research: the development and mass production of digital camera chips are attributable to the increasingly ambitious technical advances made by astronomical observation in the past 20 years.

The operation and use of these telescopes and instrumentation is not free, of course. This explains why large telescopes are usually shared by many scientific institutions, often from different countries. This is the best way to spread the costs for their design, construction, and operation. The total number of observing nights available is allocated according to the level of commitment made by each participating institution. Normally, such shares vary between 5% and 20%, which, given some 300 nights per year (365 minus bad weather and technical shutdowns for servicing the telescope and instruments), comes to about 15 to 60 nights. Every single night at an 8-m telescope, for example, then costs between $50,000 and $100,000, which has to be covered. Scientists at the partner institutions have access to the telescopes but have to apply within their own departments for observing time. In most cases there are two application deadlines per year and a several-page-long application has to be submitted to a local committee. The proposal needs to describe the scientific project and explain its importance for astronomy as a whole, the observing strategy that will be followed, and the technical details of the observations, data processing, and scientific analysis.

Since gathering new data forms the basis of observational astronomy, there are always more proposal submissions than available telescope time. Hence, demand for observing time exceeds supply many times over, particularly for the most powerful telescopes. In the end only the most promising scientific projects get their turn. This shortage especially for larger telescopes is readily explained. It is inherent to scientific progress that only the best instrumentation can tackle most of the current problems—otherwise they would have already been solved long ago. In general, large telescopes themselves are decade-long projects that develop and promote various technologies to enable the best science at the time. Even though current science always operates at the limit of what is technologically possible, one has to keep in mind that today's cutting-edge technology was actually developed some years ago. It then took years to implement and establish it. Besides the cost, the current need for more telescopes today means that more should have been built years ago. Instead, the technology for the next generation of telescopes is currently being developed.

A successful telescope time proposal usually gets the observer two or three nights of observing. It goes without saying that it is often not easy to get the telescope time in the first place. It is thus paramount to use this precious observing time as efficiently as possible. All celestial objects outside of the solar system are observable from the ground only at most for half a year because otherwise the Sun is in the way. This adds to the pressure to successfully complete the observations because if something goes wrong, new telescope time in the following year has to be won again. Although it is often possible to thoroughly prepare observations in advance, the weather is not always predictable. You are simply out of luck when the clouds refuse to dissipate, or when it rains or snows. In such cases, the observer has no alternative than to just wait for better weather and write a new proposal. Some smaller telescopes with less competition for observing time automatically add extra nights to each allocation to compensate for bad weather. Even so there are no guarantees. If technical problems arise, and unfortunately they occasionally do, the same applies. Lost time is officially entered in the observer's report, but there is usually no compensation. Tough luck!

To counteract this problem and design observations more efficiently as well as to make them less dependent on the weather, some observatories have introduced special observation strategies. The astronomers do not observe their objects on location themselves anymore but allow the telescope staff to do the job. This "queue" mode of observation, a kind of waiting-in-line strategy, makes it possible for many different observing programs to be executed exactly during the required the weather conditions. Different observations have to be carried out under different weather conditions and require different phases of the Moon and/or particular times in the night when an object is observable. For this kind of observing, it is essential that every single exposure be planned to the last detail and transmitted to the telescope operators long before they begin to observe. The planning of the observations is done by means of sophisticated software that is downloadable to any computer. Then each individual exposure is packaged as a so-called observing block, which includes all technical details and the required weather conditions and lunar phases. When everything is ready, the observer sends these instructions to the observatory and has nothing else to worry about. The same procedure applies to all observations performed by space telescopes, which are executed by the relevant control center rather than telescope personnel on site.

The queue strategy has the consequence that observers do not have to make the—often quite long—journey to a telescope. It is exciting and interesting when you have to travel to the telescopes in Chile once or twice a year. But international travel can become quite exhausting and, above all, expensive. Added to the stress for a traveler is that of the actual observing. It is dark for 12 hours in the Chilean wintertime. This means that as an observer you are not just spending 12 long hours at the telescope control room. You also have to put in a few hours in the afternoon to obtain calibration measurements and complete preparations for the observations of the coming night. All in all, you have a busy work night lasting up to 16 hours. Flying to Chile from the United States, there is practically no time shift. Since astronomers sit at their telescopes at nighttime, however, they then experience an "observer jetlag" of 12 hours. Particularly the second half of those long nights are very straining. But the prospect of new, exciting data or sometimes rather unwelcome technical problems will keep you awake and on track until dawn. Then you can drop into bed for a few hours, only to get up again soon to prepare for the next night. Besides all these efforts, there is one crucial advantage of having the observer directly at the telescope. Many decisions have to be made on the spot right after each individual exposure—not weeks ahead of time. For example, a successful discoverer of ancient stars often can then react quickly and decide whether or not it is worthwhile continuing with an observation.

7.5 Three Steps toward Success

Stellar archaeology is like searching for the needle in a haystack. In our case, the haystack is the entire halo of the Milky Way, and the needles are the rare metal-poor stars. We make use of the fact that the oldest stars carry only tiny amounts of heavy elements in them and that we can determine their chemical composition by means of spectroscopic observations. It is the only way that these few, chemically primitive stars can indeed be identified in the immense Galactic haystack.

Within 100 light-years around the Sun, its metal-rich siblings are more numerous than any of the metal-poor halo stars we are looking for by a factor of 1,000. Simplified models of the chemical evolution of the halo predict that the number of metal-poor halo stars diminishes strongly with decreasing metallicity. Stars with a tenfold iron deficiency compared to the Sun are accordingly about 10 times rarer than stars with a solar iron abundance. This implies that in the vicinity of the Sun, only about one star with less than 1/3,000th of the solar iron value can be found among the 200,000 more metal-rich halo stars. However, the farther one peers into the halo, the greater the chance of finding a metal-poor star in an equally sized sample.

If you want to identify halo stars particularly deficient in metals, you have to be able to distinguish Galactic disk stars and metal-rich halo stars from metal-poor halo stars as efficiently as possible. Otherwise you have no chance of finding these extremely rare objects from the early Universe. Hence, a systematic search requires large-scale sky surveys so that the broadest possible regions of the sky can be screened.

The search for metal-poor stars usually proceeds in three steps to gradually eliminate all the uninteresting objects. As with gold panning, you have high hopes to return home, or, better said, go to the big telescope, with something valuable. The three observational steps and the corresponding spectra with their measurable lines are depicted in Figure 7.6.

1. Coarse low resolution spectra (or narrow-band photometry) for a huge star sample are obtained within the framework of a survey covering a larger area of the sky. A relatively weak calcium K line suggests candidate metal-poor stars.

2. Higher-resolution spectra are taken of the candidates identified in the first step with 2- to 4-m telescopes. By means of the calcium K line, which can now be substantially better measured, a decision can be reached whether the star really is metal-poor. In this step, we rely on the calcium abundance being a good proxy of the iron abundance.

3. The most promising candidates from the second step are spectroscopically observed with a large telescope. These high resolution spectra enable a detailed abundance analysis of many elements.

Let us take a closer look at these three steps. The candidate selection in the first step of this lengthy search strategy is often based on wide-angle exposures for which the light initially passes through a large prism before entering into the actual telescope. With this technique, instead of obtaining point-shaped stars on the photograph, small low-resolution spectra appear at each object's position in the image. Such objective prism spectroscopy can be applied as long as the observed

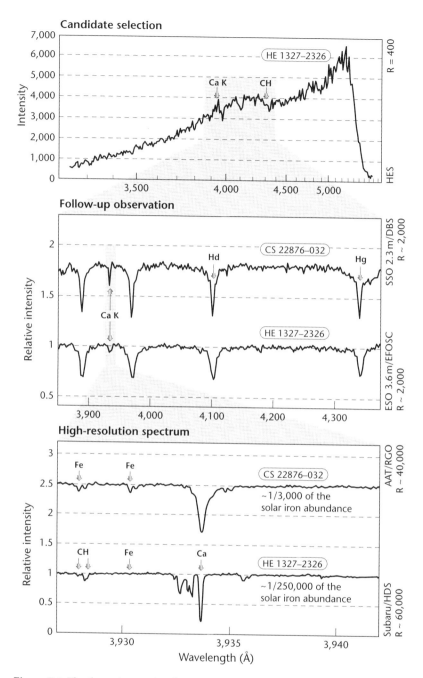

Figure 7.6. The three observational steps required to find metal-poor stars. *Top*: Survey spectra of two example metal-poor stars. *Middle*: Follow-up spectra covering the Fraunhofer lines calcium K and H at 3,933.6 Å and 3,968.4 Å, resp. *Bottom*: High resolution spectra in the range 3,900–4,000 Å. In the case of HE 1327–2326, the iron lines are no longer visible, owing to its great iron deficiency. Lines of molecular carbon appear instead because the star is very carbon-rich. All except for the upper spectrum in the top panel are normalized and, where appropriate, shifted for easy comparison. The y-axis of the upper spectrum indicates the photon counts of the spectrum. (*Source*: Peter Palm; from Frebel et al., *Proceedings of the International Astronomical Union Symposium* 228, edited by V. Hill, P. François, and F. Primas (Cambridge: Cambridge University Press, 2005), 207–212)

fields are not too densely populated regions of the sky. Otherwise, the far-too-numerous stellar and galaxy spectra would overlap. This kind of spectroscopy is relatively simple to carry out and can also be done by experienced amateur astronomers. You merely need a sufficiently large prism to cover the opening of the telescope.

The products of such a survey are low-resolution spectra of all objects in the observed region of the sky down to a particular magnitude limit. Despite their low quality, these data serve as a good stepping stone from which to continue the search. After all, at least one metal line is measurable: the extremely prominent calcium absorption line, the Fraunhofer K line, at the wavelength 3,933.6 Å. It is located just below the optical wavelength range at its blue end. Luckily, it can still be detected in relatively noisy survey spectra, even when the object indicates a relatively low content of metals making the line relatively weak. Together with the color of the star, which reflects its surface temperature, a first rough estimate of the metallicity can thus be obtained. In many cases, color measurements exist for survey stars or can be obtained from additional photometric observations.

The so-called HK Survey was the first very extensive and targeted search for metal-poor stars. It was led by the American astronomers Timothy Beers, George Preston, and Steven Shectman and was carried out during the 1980s; the survey is often referred to as the "BPS survey" after the astronomers' initials. But it was also named after the calcium H and K lines and made use of objective prism spectroscopy. In the era of the HK Survey, digital selection procedures and computer-driven analysis techniques were still a futuristic dream. Photographic plates had to be exchanged by hand in red light after each exposure during cold observing nights. The telescope also had to be focused manually to ensure that no out-of-focus spectra were produced. The analyses were performed by eye with a small hand microscope with which each spectrum was individually inspected. This way it was established whether a star had a particularly weak calcium K line. Interesting candidates were marked as such directly on the photographic plate. Most of the time there were only a handful of such stars per plate. After the coordinates of these objects were reconstructed, follow-up observations of each of them could be undertaken.

The detailed work soon paid off. For example, the HK Survey produced the first ancient halo stars whose age could be determined with the help of radioactive element decay. It also delivered a large sample of stars with metallicities down to 1/10,000th of the solar iron value and a wide variety of abundance patterns. Many of these stars are still being studied today because of their interesting chemical signatures. Based on current telescope standards they are relatively bright, which means that very high quality data could be recorded relatively quickly. Moreover, this survey proved an important point: metal-poor stars are rare, but many of them can be found in a systematic search.

Motivated by this knowledge, the Hamburg/ESO Survey was scrutinized for metal-poor stars toward the end of the 1990s. This survey had been carried out under the direction of astronomers from Hamburg Observatory, Germany, using the 1.2-m Schmidt telescope at ESO's La Silla observatory in the Chilean Atacama Desert. Its original goal had been to investigate very distant luminous galaxy centers called quasars. The resulting spectra later turned out to be a treasure trove for research on various kinds of stars.

The Hamburg/ESO Survey also used photographic plates to record the objective-prism spectra. The plates themselves were slightly larger than a vinyl record sleeve, and each one contained a field of sky covering an area of 5 × 5 degrees. For comparison, the apparent diameter of the Moon is a mere half degree. The roughly 400 fields of the Hamburg/ESO Survey contain a total of 4 million spectra with about 10,000 objects per plate. Eventually, these photographic plates were scanned for systematic inspection and analysis, not by a microscope but with computer algorithms. In addition, color information was available for each star, making the search for any metal-poor stars among the 4 million spectra considerably simpler and more efficient.

Naturally, a sample as huge as that of the Hamburg/ESO Survey contains numerous galaxies, quasars, and other exotic objects besides all the stars. Using suitable search algorithms, the first selection of metal-poor candidates can still be undertaken. The aim is to distinguish between different stars and to roughly recognize whether the calcium K line is particularly weak with respect to their color and surface temperature. The noisy low resolution of these spectra is just as much of a challenge for computers as for human experts, though. As a result, the list of selected candidates did not exclusively contain true metal-poor stars but various false positives as well. Consequently, visual inspections of the long candidate lists from this first step were necessary.

For my doctoral thesis, I initially received such a preselected sample of 5,500 of the brighter candidates. My first task was to closely inspect each spectrum to establish if it was a good metal-poor candidate. The candidate selection of my sample is described in more detail in chapter 10. This visual inspection led to the identification of roughly 3,700 false positives.

There was one major difference between the Hamburg/ESO Survey and the previous HK Survey. The Hamburg/ESO stars were all substantially fainter, which had consequences for the follow-up observations. Observations of fainter stars take considerably longer. Constraints on the limited telescope time prevented follow-up work for many of these stars. Furthermore, the Hamburg/ESO Survey produced a lot of candidates. From the over 7,000 low-resolution candidates selected by means of computer algorithms, only about 2,500 stars have been observed with medium-resolution for more precise calcium-line measurements. Most of the faint stars still await observations, but as they require relatively long exposure times, it is likely that they will remain unobserved. These stars, if they ever were to be examined with high-resolution spectroscopy, would necessitate a lot of time with a large telescope, which is very costly. Discarding the faintest stars is often a compromise that has to be made, even when it will never be known if any potentially interesting stars were lost this way.

More precise measurement of the calcium K line calls for medium-resolution spectra to be taken by telescopes equipped with mirrors from 2 to 4 m in diameter. Independent of its actual element abundance, iron, owing to its physical atomic properties, has much less prominent absorption lines than the strong resonance line of calcium, for example— the calcium K line. Accordingly, in the second step, weak iron lines still remain hidden in the noise. This is especially true for the weak iron lines in the spectra of the most metal-poor stars.

The high-resolution spectra of the third and last observational step make it possible to finally determine how deficient in iron the star in fact is. In order to obtain such high-resolution spectra, the starlight has to be dispersed extremely broadly. This enables even very weak spectral lines to become measurable. To capture enough photons at each wavelength, these observations require the largest optical telescopes in the world with primary mirror diameters of 6 to 10 meters. As already mentioned, observing time at those telescopes is expensive and hard to come by— usually no more than a few nights per year. We will thus never be able to observe all the candidates. Only the most promising, most metal-poor candidates can be looked at in this final round of gathering spectroscopic data.

Comprehensive chemical analyses of stars can be performed only with such high-resolution spectra. Quite a few lines of other elements besides iron are identifiable, such as carbon, sodium, magnesium, titanium, nickel, strontium, and barium. After all these absorption lines have been carefully measured for their strengths, the corresponding element abundances can be calculated with the aid of computer-simulated stellar atmospheres and then interpreted against the backdrop of chemical evolution.

The high-resolution spectroscopic observations of the best and presumably most metal-poor Hamburg/ESO stars provided the research field of stellar archaeology some new highlights: The first star with a then record-breaking iron value of just 1/150,000th was found. It was absolutely a sensation. Many extremely metal-deficient stars gradually joined it, among them a whole series of rare, metal-poor stars whose ages could be determined from the observed abundance of radioactive thorium. The crowning success of this survey was my discovery of the second record-breaking star with 1/250,000th of the solar iron value and of a star that could be dated with radioactive uranium and thorium to an age of 13 billion years.

7.6 Observations with MIKE

When observing with high-resolution spectrographs at large telescopes, many different things have to be kept in mind and carefully considered. Some decisions can be made prior to the observing run, but some can be made only just before pressing the "start exposure" button. A few of these considerations are introduced here in some detail. Our observations of a dwarf galaxy star with the MIKE spectrograph ("Magellan

Inamori Kyocera Echelle") at the Magellan Clay Telescope is used as an example.

A limiting magnitude exists for each kind of observation. It describes up to what brightness (actually, faintness) an object can still be successfully observed, in terms of technical feasibility. For high-resolution spectroscopy, this limit is currently at a visual magnitude of about $V = 19^m$.

For stars considerably brighter than this limiting magnitude, the quality of the spectrum is simply driven by the total observing time. The fainter the object, the longer the exposure times have to be. Longer exposures are, however, often broken up into sub-exposures. Exposures can be repeated to build up the signal in the spectrum as long as there is enough telescope time available. Fainter stars closer to the limit can use up large chunks of telescope time, though, since more than a few hours per star would be required to obtain a spectrum with even a low signal-to-noise ratio. We thus limit our observations to metal-poor halo stars with visual magnitudes between $V = 12^m$ and 16^m. These stars are 600 to 15 times brighter than the limiting value.

When observing stars at the limiting magnitude, various aspects suddenly have to be taken into account to ensure that a useable spectrum is actually being obtained. Faint stars of $V = 19^m$ require very long exposure times of up to 10 hours to build up a signal-to-noise ratio just barely useful for analysis.

Compared to halo stars, when dealing with stars in dwarf galaxies, only few bright stars are available. Dwarf galaxy stars are obviously all located inside their native dwarf galaxy, far beyond the Galactic halo. Accordingly, the stars appear very faint in the sky even when they are intrinsically fairly luminous giants. But those giants are the only stars we can barely observe with telescopes equipped with mirrors of 6 to 10 m in diameter.

When the sky is not completely clear or the air is too turbulent, these already lengthy observations take even longer. The so-called seeing describes how much of the stellar light is blurred by air turbulence and is usually measured in arcseconds. Depending on the variation in air quality, the star can appear smaller or larger in size. Smaller is considerably better because the light falls onto the spectrograph in a more concentrated beam. When bad seeing causes the star image to become larger than the width of the spectrograph slit, star photons are wasted—as if you were trying to fill sand from too large of a funnel (the starlight) into a bottle with too narrow of a neck (the slit). If, however, the funnel has a much narrower opening than the bottleneck, the filling procedure is inefficient as well.

All in all, this means that an observer of very faint stars is particularly dependent on the weather. The preference is, of course, if possible, not to lose any of those already few photons. To ensure a useful result despite bad weather, the observer can adapt the slit size to the conditions. A larger slit reduces the quality of the data, that is, the spectral resolution. But often this compromise is better than not getting any data at all. Having started to observe an object with a particular slit width it should best not be changed anymore, even if the weather later improves.

There is yet another significant problem for hour-long exposure times: cosmic rays. Cosmic rays are high energy particles of cosmic origin that are constantly permeating everything, and are therefore also bombarding the CCD detectors. This bombardment is harmless to us but not to a detector pixel. If a cosmic ray is registered by a pixel during an exposure, there is no chance left for that pixel to properly record photons from the faint star anymore. For this reason, long total observing times have to be divided into many shorter sub-exposure. Individual observations of 20 to 30 minutes are best.

But once again, for such faint objects as those with $V \sim 19^m$, another problem related to such short exposures inevitably crops up. Only after the detectors has registered a certain number of stellar photons in all their pixels, the intrinsic noise level of the CCD chip is overcome. Otherwise, no useful spectrum (i.e., stellar signal) can develop on the chip. With bright stars the detector noise is overcome shortly after the exposure has begun. These very faint stars, however, require observing times of more than 30 minutes for any individual exposure to do so. Not nearly enough stellar photons reach the detector in a shorter amount of time. Exposure times of one hour just about suffice to collect enough photons to produce an actual spectrum above the detector noise. Within an hour, however, one also collects a significant number of cosmic rays, but the damage to the stellar spectrum is still acceptable. Thus, every single exposure is a compromise. These technical details ultimately determine where the real limit for high-resolution spectroscopic observations of faint stars lies.

In addition, the chosen wavelength range of the observation plays an important role as well as the decision about which spectral resolution to choose. Both of these issues directly determine how many photons of a star can be collected. Experience shows that one should not have too high expectations. With such faint stars, the spectrum is usually useless below 4,000 Å, even at a total exposure time of 10 hours. These red giants stars do not shine as much in the blue spectral range compared to the red. Furthermore, the CCD is less sensitive in the blue, which amplifies the issue that blue photons are difficult to collect. Such long and somewhat risky observations are therefore advisable only for particularly important stars. One example is an extremely metal-poor star at [Fe/H] = −3.8 in the Sculptor dwarf galaxy that I observed with MIKE in July 2009 at the Magellan Clay Telescope.

This red giant star, S1020549, was observed for a total of 8 hour and 20 minutes. An excerpt from my observation log in Table 7.2 shows the entries of the individual exposures including the file, name, universal time (UT) at the beginning of each exposure, exposure time (t_{exp}), an indicator of how much airmass is present in the direction of the star, the seeing, and spectrograph slit width used. Bias, quartz-flat, and milky-flat frames are required to characterize the noise distributions and properties of the CCD chip, to subtract them later from the actual stellar observations. Quartz-flat frames are needed to accurately trace the complete spectrum on the chip. This way, the spectrum can be extracted and processed with great precision because in the end we want to retain just the starlight's contribution to the exposure and nothing of the sky background.

The "airmass" is larger when a star on the sky is closer to the horizon. The light of a star directly above the telescope has to traverse the least amount of airmass. How the star gradually rises and sets can thus be followed with the variation in airmass. Every night, I observed S1020549 as soon as possible after it had risen. Figure 7.7 illustrates the course of the observations on the sky by the two Magellan Telescopes during one of those nights. The separate, clustered points of the Clay Telescope entries reveal that throughout that night I observed many halo stars with shorter exposure times in addition to the faint dwarf galaxy star.

Table 7.2. Log Book Entries from the Two Nights during which the Metal-Poor Sculptor Star S1020549 Was Observed with the 6.5-m Magellan Clay Telescope

File	Name	UT	Exposure time (t_{exp})	Airmass	Seeing	Slit	Comment
1001	Bias		0 s			—	Afternoon
to	Quartz flats		13 s			0".7	calibrations
1032	Milky flats		50 s			0".7	
...							
1142	S1020549	5:37	2,400 s	1.5	0".6	0".7	
1143	S1020549	6:18	2,400 s	1.3	0".7	0".7	
1144	S1020549	6:59	2,700 s	1.2	0".6	0".7	Clouds ...
1145	S1020549	7:45	2,700 s	1.1	0".5	0".7	More clouds
...							

File	Name	UT	t_{exp}	Airmass	Seeing	Slit	Comment
2001	Bias		0 s			–	Afternoon
to	Quartz flats		13 s			0".7	calibrations
2032	Milky flats		50 s			0".7	
...							
2096	S1020549	5:43	2.400 s	1.4	1".0	0".7	
2097	S1020549	6:24	2,400 s	1.3	0".8	0".7	
2098	S1020549	7:04	2,400 s	1.2	1".0	0".7	
2099	S1020549	7:45	2,700 s	1.1	0".8	0".7	
2100	S1020549	8:31	2,700 s	1.0	1".0	0".7	Bad seeing
2101	S1020549	9:16	2,700 s	1.0	1".3	0".7	Seeing!!
2102	ThAr lamp	10:02	4 s			0".7	Calibration
...							

These individual observations lasting over the course of two nights were worth it: Our chemical analysis of star S1020549, together with the interpretation that dwarf galaxies such as Sculptor have contributed in the past toward the assembly of the Milky Way's halo, was published in the international scientific journal *Nature* in the spring of 2010. And S1020549 was the most metal-poor star of any of the dwarf galaxies

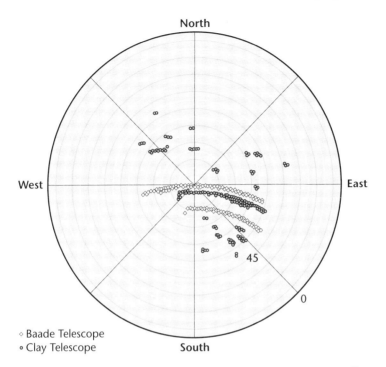

Figure 7.7. Nightly observations across the sky as carried out by the two Magellan Telescopes. Adjacent points indicate longer observing times during which the object was followed on the sky, in some cases for hours. (*Source*: Peter Palm; reproduction of measurements from *Magellan Telescopes—Guide Camera Seeing and Sky Trace*, Chile: Las Campanas Observatory)

until the middle of 2010. Then colleagues of mine found a dwarf galaxy star that was even more deficient in metals, which ultimately confirmed that classical dwarf galaxies do possess extremely metal-poor stars and that they are well suited for helping us reconstruct the history of the formation of the Milky Way.

CHAPTER 8

LET'S GO OBSERVE SOME STARS!

Astronomers have been using telescopes for 400 years to examine cosmic objects in the night sky. What once started in the 17th century with Galileo and small lenses made from polished glass is now one of the most technologically advanced enterprises in science. With time, bigger and bigger mirrors have been cast to suit telescope designs of ever growing dimensions because it is the only way to capture enough light of increasingly faint stars and galaxies. The grand aim has always been to keep forging ahead, deeper and deeper into the space to explore the Universe.

In this chapter I share a number of stories based on my practical experience as an observer. After all, the daily (and nightly) routines when observing at a professional telescope are anything but boring. Astronomers have to, among other things, take into account the forces of nature, such as rain, wind, and fire, and even such profane matters as a shortage of drinks.

8.1 Going Stargazing

As an astronomy student at Mt. Stromlo Observatory in Australia and working on an observational project, I had to go on observing runs several times a year. The university's own 2.3-m telescope is part of Siding Spring Observatory, which is located in the northern part of the Australian state of New South Wales. It is nearly 600 km from Canberra. Before my very first run there, I had been keen to see a real professional telescope. The thought of using it by myself was very exciting. I had received some basic instructions on how to operate the telescope, and my observing run would last for 10 nights.

The various telescopes of this observatory are situated on a small mountain in the middle of the pretty Warrumbungle National Park, which is home to countless kangaroos and other inhabitants of the Australian bush. The park is named after the surrounding mountain range. At its eastern end lies the small town Coonabarabran with its 2,500 residents. There is a small main street with a large clock tower in the center of town and a few small stores where the bare essentials, such as beer, bread, and milk, can be purchased. Coonabarabran is thus not much more than a sleepy little village. But there is one very special thing about it: it is the self-proclaimed Astronomy Capital of Australia. The name Coonabarabran derives from the local Aboriginal word for "inquisitive person," which suits an astronomy capital and more general Australian ambitions to conduct good astronomy very well. "Coona" is only half an hour away from Siding Spring Observatory. Tourists can visit it in the daytime and view the telescopes. They are welcomed by a big sign, as can be seen in Plate 8.A.

At the observatory, observers are accommodated in small, simple rooms. The daytime staff cook delicious lunches and dinners and prepare the "night lunch"—the midnight snack—which observers take to the telescope. Usually, there are four to six observers at the various telescopes. They dine together in the evening, discussing the successes and failures of the previous night, the weather forecast for the next one, technical details, and astronomical or other news.

On the mountain, there is always a gorgeous view onto the national park below. If one cannot or does not want to sleep during the day despite observing all night long, one can go on nice walks through the Australian bush, for instance, to the 90-m "Breadknife." It is a quite steeply protruding rock made of solidified lava that can be reached by foot in about an hour's walk through eucalyptus trees. I made this tour once when the weather was cloudy and unsuitable for any observing.

The building containing the telescope I was using at that time is as large as a multi-story house. It can be seen in Plate 8.B. The entrance is on the ground floor, where various fuse boxes and large amounts of cables are coming down from beneath the telescope mirror on the floor above. The bottom part of the telescope, with its 2.3-m mirror, is located on the second floor along with several instruments for various kinds of observations that can be individually attached to the telescope as needed. The control room is located on the third floor, which is halfway up above the telescope and instrument platform. Since the telescope is built on a slope, there is a partial basement level below the entire telescope building. While most telescope buildings have a round rotating dome, the entire building of the 2.3-m telescope rotates when slewing from one object to the next—apart from the basement. A small kitchen and a toilet are located there.

To visit the bathroom facilities during the night there are two ways to get there. One route goes inside the dark telescope building down to the ground floor and from there to a small trapdoor that leads into the basement below the telescope. Using a flashlight on the way down there is not a good option because the observations would be adversely affected by the light.

Once, in the early morning hours, when I made my way underneath the telescope and to the stairs going into the basement, I did not realize that the staircase was equipped with a light barrier for safety reasons. When the large motors that move the entire building are directly above the staircase, any crossing of that light barrier triggers an alarm. One has to indeed watch out not to bump one's head, especially all alone and with no one there to help should an injury occur. A short while later, as I was carefully making my way back up again, a screeching alarm made my heart skip a beat! No one had told me what would happen if this alarm ever went off. My first thought was that something terrible had happened to the telescope and that surely I must have caused it. And that it had to happen precisely at that unfortunate moment when I was not even in the control room but in the basement! I could already see myself being held accountable for expensive and time-consuming repairs. My panicking lasted only for a few seconds, but another half hour went by before I could relax again, only after I had conducted various checks to verify that the telescope and the instruments had not been damaged. I certainly was not tired anymore that night.

The other route down to the basement leads via an external metal-grid stairway from the control room on the third floor, halfway around the telescope building. Because of my experience with the light barrier, I preferred this route via the outside stairs. Then I could take a glimpse at the weather at the same time and see where the Moon was in the sky.

The outside stairs were not always very pleasant either, though, especially in bad or cold weather. Often, the wind was blowing so hard that I had to hold on tight to the railing while clutching the flashlight. This route had its surprises, too. I was not prepared to suddenly see two, then four, then six brightly flashing eyes appear in the beam of my flashlight directly in front of me at the end of the stairway. They belonged to some grazing kangaroos peacefully munching away. When a kangaroo gets up on its hind legs, its eyes are at about your own eye level. I had apparently disturbed their meal time. But I sure jumped up like a kangaroo myself that time, absolutely frightened.

Kangaroos can often be seen at Australian observatories, during the day as well as at night. But there are also quite a few other smaller creatures. Once, during a longer exposure, I wanted to open the door to the basement coming from the outside. Just at that moment an enormous toad started croaking loudly, right at my feet. Various spiders, millipedes, and other creepy-crawlies that were busily skittering about on the basement floor, as well as that stale, foul old-basement smell, left their own impressions, just to round off every visit there.

Sometimes it really was a bit spooky at the 2.3-m telescope. After all, it is pitch-dark outside, you are all by yourself, and there are all sorts of unfamiliar noises. The wind often howls, rattling the door to the outside and the entire telescope building. In addition, you get sleepier and sleepier the later it gets at night. But of course you are not allowed to fall asleep because the telescope has to be operated and stars must be observed. You are nicely shaken when the telescope slews from one object in the sky to the next or if during a long exposure the building position is slightly adjusted to compensate for the Earth's rotation. The heavy motors of the telescope building can make some sudden and quite eerie noises. In the middle of the night, feeling tired, I was regularly startled and outright frightened, and almost fell off my chair because each time it sounded as if someone was trying to break in. But the trusty telescope was only doing its job.

Another student, also learning and observing, was meant to join me at Siding Spring a few nights later. So I spent the first three nights there on my own. Plate 8.C shows the telescope itself with the spectrograph I used. Smaller telescopes such as this 2.3-m telescope are serviced by the technicians only during the day. Consequently, there is no assistance available while observing. That meant the whole night I was sitting alone and completely left to my own devices in the telescope control room somewhere in the Australian bush.

After my fellow student arrived at the observatory, the first thing I told him about was the terrible alarm. I certainly did not want to hear that sound again. After I had completed my report on the state of the telescope, its instruments and alarms, and the musty, creepy-crawly-filled basement, we quickly came to the conclusion that we actually ought to be writing a mystery novel. An observatory would be very well suited as the crime scene for some dramatic plot. At that early hour of 4 a.m. and a little overtired, our imaginations ran wild. He then had the fabulous idea of creeping up on me from behind in the dark to pounce on me with a loud "Boo!" That was not quite amusing. Despite all this early morning kidding around, we were good and conscientious observers. We were always checking the sky outside. Is it clear? Are there any high-altitude clouds? Where is the Moon compared to where the telescope is pointed? Everything was in the best of order, and we took many spectra that night.

The rotating boxy type of building of the 2.3-m telescope makes it possible to climb directly onto the building's roof and to watch the world from above—day and night. Notwithstanding all the minor or major adventures one inadvertently experienced during an observing run, it always was something special to simply lie flat on your back up there on the telescope roof and admire with your own eyes the twinkling lights of the Milky Way. As it is pitch-dark there, I could experience the Southern Hemisphere celestial sky and the Milky Way right above me in its grandeur for the first time. Up there on the roof I felt much closer to the stars and the sky.

As the morning approached and the last observations were complete, the telescope and its building needed to be rotated back into park position. Everything had been rotating back and forth all night. A full 360-degree rotation of the telescope building took several minutes. I often used this time to sit on the external stairway. Or I would run directly up onto the roof to let myself be driven into the sunrise by the rotating telescope. That roof is unfortunately inaccessible these days, for safety reasons. In my days, though, I did enjoy being able to take in the fresh morning air, stretch out after a busy ten-hour work night, and enjoy the colorful display of the rising Sun.

8.2 Good-Weather Beer

Some observatories allow astronomers to bring along beer or wine for dinner. This was the case at Siding Spring Observatory, where I carried out seven different observing runs for my PhD thesis research. My fellow student who was now there for his own observing run at another telescope had brought along a six-pack of beer, which normally lasts about a week. And as it happened—after five days, there was one beer left in the fridge.

The weather had been quite good all week long and there was just one night left to observe. It is always disappointing to end an observing run with a night of bad weather. We were eagerly hoping for one more clear night. Indeed, it looked quite promising that afternoon. Later that evening my colleague then decided to drink his very last beer at dinner. Soon it was discovered that the beer was not in the communal fridge anymore. Someone had absconded with the final beer—a rather scandalous occurrence! Night fell and we both began our observations in good spirits. But soon thereafter the wind turned and a thick layer of clouds rapidly covered the entire sky. Unfortunately, this put an end to our observations.

In the case of bad weather, such as clouds, rain, or even a strong wind, the observer simply has to wait it out. The only thing to do is wait and hope that the situation soon improves, though this can take several nights. With the observatory's "all-sky" camera it is possible to monitor the weather outside. One can then directly watch the thick clouds "live" above the telescope. Alternatively, one can get frustrated by watching satellite images of clouds and bad weather forecasts on the Internet. Poor weather is simply bad luck. Nothing can be done about it, aside from reapplying for new telescope time a year later.

In answer to the question of how long one should stay at the telescope in bad weather, my PhD advisor told me the following rule of thumb: It depends on the size of the telescope. At a telescope with a mirror 2 m in diameter, you can go to bed at 2 a.m. if the weather looks more or less hopeless. At a 4-m telescope, you may close up shop only at 4 a.m. Consequently, with 8-m and 10-m telescopes you are not allowed to sleep at all. That, evidently, is when this rule breaks down. As an ambitious student, I often opted to not stick to this rule, though. I stayed up considerably longer than 2 a.m. (conforming with the 2.3 meter telescope) with the help of loud music. I could always catch up on sleep at home. I was often rewarded for this. About an hour before sunrise, around 4 a.m., during a summer in Australia, it often cleared up. Then I could observe some more stars. I experienced this weather pattern repeatedly at Siding Spring Observatory. In some cases that was the only hour of the entire night I could observe.

If you have to sit around for hours on end, waiting for better weather, you have a lot of time to kill. You can either work or surf the Internet or pick your nose. Apart from the kangaroos, there is not much entertainment. If you are observing by yourself it can get pretty lonely, and nights can seem very long. If you are observing with someone else you can at least keep each other entertained. Around 3 a.m., such conversations may well become rather philosophical. On that cloudy last night, that colleague of mine dropped by to see me at "my" telescope, as he was as bored as I was. We complained about the terrible weather until a thought struck him. The weather must have suddenly turned for the worse only because somebody else had finished off his last beer! Of course! There could not possibly be any other reason! Not long afterward we postulated the theory that any good observer should always take a "good weather beer" along on any observing run to be drunk by only the observer before the last night. That would guarantee good weather for the whole run—unlike what happened in our case, when for inexplicable reasons the weather turned bad only for the very last night.

8.3 A Sunset

In the summertime, nights are shorter. This means less stress and more sleep, but of course also less observing time. It is also warmer and more pleasant to be outdoors and to occasionally soak up some evening sunshine. After all those busy afternoon preparations for the observing and the final tests just before dark, a very special moment comes. It is very brief and yet hugely impressive: the sunset. In Chile, at Las Campanas Observatory, the observers and staff of the various telescopes come together to watch the Sun go down. Shortly before the Sun touches the horizon, you can already clearly feel it getting cooler and the wind often picks up a little. You might be standing outside on the catwalk right by the telescope building to quietly watch how the day turns into night. As most telescopes are situated on mountains, the view is always spectacular and breathtaking, especially when the whole world, slowly but surely, first takes on a tinge of yellow, then orange, and finally red light. Shadows stretch longer and longer as the Sun gradually sinks down. During that moment each one of us is truly an observer, an observer of the Sun, of the world, and of oneself. It is a brief moment in which it seems the world has come to a standstill. Everything around you is calm and peaceful. You can take the time to breathe deeply and to feel the gentle power of being out there. You are suddenly part of nature, and you yourself are not that important anymore. You are at one with everything around you. The grand admiration I feel during these moments immediately compensates for all the stress of traveling and from having to work long nights.

Shortly before the Sun disappears completely the very last bit of the upper rim of the solar disk is still visible above the horizon. With some luck you then can see a so-called green flash. As is shown in Plate 8.D, the last small yellow bit of Sun just above the horizon will glow bright green for a fraction of a second. For this brief instant it looks like a flashing green diamond. Almost as if it were a thank-you for having made the effort to come and watch the sunset. Unfortunately, this thank-you is pretty rare, as a green flash is visible only if the air in the Sun's direction is particularly clear, and preferably if the Sun itself is setting behind a tall mountain at the horizon. Even to astronomers a green flash is something exceptional, despite the fact that many telescopes are located on mountaintops that generally offer stunning views.

Green flashes are attributable to light refraction. Sunlight is refracted in the atmosphere in the same way that light is refracted when it hits a water surface at an angle. The velocity of sunlight in our atmosphere is slightly lower than in space, the optimal vacuum. The degree of decrease depends on the density of the atmosphere. At the horizon the atmosphere is somewhat denser than in layers higher up. For that reason all rays of light that reach an observer have been bent a little bit upon entry into the atmosphere, light near the horizon somewhat more so than light coming directly from overhead. In addition, blue and green light, with its shorter wavelengths, will bend more than red light.

When the Sun has almost completely set, the horizon covers not only most of the solar disk but also all the red, longer wavelengths of light because they are less bent. The red light simply does not make it past the horizon anymore. But in that brief moment, first the green light still manages to do so, but in the end only the blue light is bent enough. This is exactly what is observed as a "green flash" and even a "blue flash." Consequently, a "blue flash" is visible after a green one, as soon as the green light has also disappeared behind the horizon. Blue light is also generally dimmed by the atmosphere, making a "blue flash" an extremely rare event. But some of my fellow observers have indeed seen "blue flashes" in Chile. I guess I have to do much more observing before I can witness this natural phenomenon myself. But I did once see a "green flash." It was during my very first observing run in Chile. It was very exciting and of course became one of the discussion topics at dinner the following day.

The ritual of watching the sunset and hoping to be lucky enough to catch a "green flash" happens every evening. It is a wonderful habit because you can briefly enjoy the calm before the storm. At the same time, you are outside and can check whether there are any clouds in the sky or whether the air is hazy or anything else looks suspicious. Although you cannot do anything about the weather, at least you can roughly predict how the night will turn out. And, of course, you always hope for a night without any technical problems.

The Sun vanishes completely behind the horizon, as shown in Plate 8.E. You take another close look to make sure it is time to get started. Soon you can see the first bright stars twinkling above. This means that the final preparations for observing need to be made. It is still light outside, but the time up to the end of the so-called astronomical twilight passes swiftly. You draw the heavy curtains of all windows in the control room, stick your head out the door one last time, usually let it shut with a loud noise, and then get to work. I am always itching to get started right away, especially on the first night. A good astronomer, after all, wants to use every single minute of darkness for observing and not waste any time.

Once observing has begun you quickly resurface from this mystical world between heaven and Earth that had enveloped everything in rose-colored hues. A long and busy night lies ahead during which a little bit more of the Universe is brought down to Earth.

8.4 The Observa-thon

Observing puts you into a strange state of consciousness in which you are neither asleep nor awake. It is a kind of daydreaming because you experience everything very intensely at the given moment. But just a little later on you can hardly remember the details anymore because you lose your proper sense of time. The long nights are spent selecting the next targets, setting the exposures on the computer, and immediately inspecting the spectra using various software programs. Then you either stay with the same star for a while or move on to the next. This goes on, hour after hour, until dawn, disrupted only by the occasional trip to the small kitchen downstairs or a triumphant "yeah!" when a new exposure shows a high level of photons and thus a nice spectrum. But then you have to immediately deal with the next exposure.

These procedures are independent of the telescope you happen to be working with. The sole difference is that at larger telescopes trained staff operate the telescope. The observer has to operate just the instrument and of course make all the decisions regarding the observing.

A regular night of observing in the Chilean wintertime at the 6.5-m Magellan Telescope goes roughly like this:

15:00 Time to wake up. . . . The first thing I do is to stick my head out the door to check the weather. Hopefully the sky is clear!

16:00 Short drive up to the telescope. Various calibration measurements and comparison lamp spectra need to be taken for subsequent data processing, all before the onset of darkness.

17:00 The calibrations are running; time to check email and look forward to dinner.

17:30 Dinner with the other observers! It is the best part about going observing, except, of course, good weather conditions and fantastic new data.

18:15 Quick, quick, back to the telescope. It is slowly beginning to get dark.

18:20 Sunset. In wintertime there is usually not enough time to watch the entire sunset. Some target stars have not yet been selected and a rough schedule for the night needs to be drawn up. Then it gets a bit hectic before the first observations can begin.

18:40 The telescope slews to the first target star. It is not yet completely dark, but very bright stars that function as comparison objects are already observable.

19:00 Any subsequent observations are moving along quite smoothly now. But it is only 19:00—only another 12 hours left! Best not to think about it. . . .

21:15 Time for a first cup of Milo, a delicious malt-flavored hot chocolate. From experience I know that I need the sugar to stay awake! I am no coffee or coke drinker.

22:00 It is considerably colder today than yesterday. The cold air currents have caused some turbulence in the atmosphere. The "seeing" is fluctuating a lot and not particularly good. That is annoying because now all my observations are going to take twice or three times as long—and even longer if conditions continue to deteriorate.

24:00 The seeing is still pretty mediocre but some photons are reaching the detector. What else do you want?! More photons, of course. But faster, please!!

2:00 "Night lunch" time. For tonight I ordered two homemade whole-grain bread sandwiches filled with avocado, red peppers, tomatoes, and lettuce, with a dash of hot and spicy Chilean tomato ketchup. That will be nice and refreshing. Add to that another cup of Milo.

3:00 My exposures seem to be yielding hardly any photons by now. Or am I just imagining this? One is a bit tired at 3 a.m., you know. . . .

4:00 It is cloudy, no wonder. None of those stellar photons can get through to us. How disappointing. Nothing is worse than having clouds above the telescope. Then again, rain, a storm, or snow would clearly be even worse. But you have to keep thinking positively: the weather has often been very poor of late, so it should be clearing up again soon. It is not particularly pleasant in the control room either. The humidity is at 8%, whereas outside it is only 5%, making me feel like a dried raisin.

6:20 The last exposure. Quick, quick, it is already getting light outside.

6:30 All done. The night is over. Or rather, not quite yet.

6:35 What would I give just to be able to nod off right here and now?! But, alas, the calibration exposures of the not-yet-too-bright twilight sky still need to be taken. There is no way around it.

7:05 The twilight sky spectra are done. At last.

7:10 Now I am really done and exhausted.

7:20 Back in my room. Quickly shutting the heavy curtains so that it stays nice and dark in my room. And it's off to bed.

7:35 Sunrise. Outside. On the other side of my curtains and totally unimportant to me right now because . . .

7:45 Zzz zzzz zzz zzzz. . . .

8.5 One Hundred and Five Stars per Night

For my PhD thesis research at the Australian National University, I spent a total of 42 nights spread over two years, carrying out observations with the 2.3-m telescope at Siding Spring Observatory. The shortest observing run took 3 nights, the longest was 12 nights. The former is almost too short because you do not have enough time to adjust from the daytime schedule to the nighttime routine; the latter is pretty long because after one week you are beginning to feel like a vampire. Darkness has become quite comfortable and when you wake up daylight seems quite glaring. You have not exchanged a word with anyone for days on end, apart from brief dinnertime chitchat. You live quite encapsulated in your small "observer's world" at the telescope.

My star sample consisted of 1,777 stars, of which I observed about 1,250 during those 42 nights using the "double-beam" spectrograph at the telescope. To assist me, the remaining 500 stars were observed by a few of my colleagues also with the 2.3-m and similar telescopes in Chile. My thesis advisor also helped out. One time he returned from observing and told me that he had observed 95 stars in a single night. Up to that point I had never counted my nightly rate. It changed from that day forward. For my next run there was just one goal: to observe more than 95 stars in a single night! I could not bear that my advisor would hold the record having worked on "my" stars.

I did, in fact, later manage to observe over 100 stars per night during several nights. My new record of 105 stars had been based on the following strategy. You need as long and clear a night as possible, many very bright stars that are located as close to each other in the sky as possible as it reduces the slewing and changeover time of the telescope going from one object to the next, and good loud music to prevent you from getting sleepy. The observations then generally proceed as follows: You slew the telescope to the position of the first star. With the help of a smaller camera attached to the front of the telescope, an image is taken of the region of the sky in which the to-be-observed object is located. That way you can test whether the star is correctly positioned for the observation. If it is not the case, various corrections can be applied. The observing times for my stars at the 2.3-m telescope range between 20 seconds and 5 minutes, depending on the magnitude of the star. Once the exposure is completed, you wait until the data—the stellar photon counts—are read out from the detector. This takes between 10 and 15 seconds. In the process, the "observation" is converted into an electronically readable file. In my case, of course, these are spectra and not photometric images of stars. After all, I am a spectroscopist. Then, a brief calibration measurement needs to be performed in which the spectrum of a comparison lamp filled with a known noble gas is taken. Usually, a thorium-argon vapor discharge lamp is used for this purpose. These calibration measurements enable identification of the wavelength scale of the star's observation. This takes another 30 seconds. Then it is off to the next star.

Throughout this entire procedure you are dealing with various screens, keyboards, buttons, and switches. At smaller telescopes such as the 2.3-m one, there is just one person responsible for operating everything: me, the observer. Strictly speaking, as an observer you are not just an astronomer and observer but also telescope operator, instrument specialist, and, of course, expert in all malfunctions, breakdowns, and other problems that can arise when working with a telescope.

At the same time you have to decide which object is strategically the best to observe next. In a given ten-hour night, I observed on average one new star every 6 minutes. All those different intermediary steps keep you occupied, leaving no time to get bored. On the contrary, the fainter stars requiring exposure times of 5 long minutes always offered welcome breaks for a quick breather, stretch, and change of music.

I never did check whether my advisor had really observed 95 stars in one night. But one thing is clear to me—we both are very ambitious observers, and he surely wanted to challenge me with his record. It worked. In the process I devised my own strategy to plan my observations with the maximum efficiency and to get all my 1,250 stars observed as fast as possible.

8.6 Computers, Computers . . .

Successful observations depend not just on clear weather. One night, when my colleague was visiting me again at the 2.3-m telescope and I was in the process of checking star after star off my target list, he inadvertently pressed a random button somewhere on the telescope console. We never found out which button it had been and what exactly had happened. But one thing was certain right away: the computer that operates everything was, at the very least, deeply offended and decided to conk out. Nothing, absolutely nothing worked anymore! And this had to happen when the weather was clear and the night was young.

In those first moments, as I was later told, my face turned slightly green. It was immediately obvious to me that such a breakdown meant losing an entire night to technical problems. But that was the least of my problems compared to my lot in history: the person who had completely trashed the telescope! What a terrifying thought.

Despite the mess you do need to know how to get out of such tricky situations as quickly as possible, particularly when you are sitting all by yourself at the telescope and especially when all you really want to do is to continue observing. In moments like these you learn a great deal about the telescope, beyond the scientific and technical details you normally need to know to observe, including how you can fix a device or instrument that is causing (repeat) problems or even fails completely. It is particularly fun when you do not have the faintest clue what is broken or what the solution could possibly be. Then you simply have to get creative. In most cases, the strong urge to continue observing usually provided me the necessary creativity. There is no real alternative anyway. I always managed to eliminate the problems in the end, even if it did sometimes require quite some trials that involved nervous fiddling around and running up and down stairs.

One time the telescope was hit by lightning. After a tremendous boom, it went "bsssssst" and then everything went dark and quiet. Luckily, I remembered that someone happened to mention at dinner in the observers' lodge that every telescope is equipped with a backup generator. Eventually, armed with my flashlight after having frantically searched for it a few minutes in the dark, I found the small storage closet at the rear end of the control room. I did indeed find something that looked like a generator. Not .that I had ever operated a generator before. But buttons are meant to be pushed. Soon it went "bsssssst" again and everything started back up. The power supply did, at least. A few computers and software programs in the control room and a particular camera down below on the telescope platform had to be restarted. But I was already familiar with that procedure from other problem-solving sessions. After that, everything was working again.

But back to the story of the dead telescope computer. It happened around 9:30 p.m., and technical assistance in emergencies could only be obtained by phone until 10:00. However, observers were advised to call after hours only with really important problems. Speed was therefore of the essence. I had to find out as quickly as possible whether I could somehow persuade the old VAX computer from the 1980s to start working again. My colleague began to help me by hectically digging out operating manuals and instructions from drawers and binders hitherto unknown to me. Meanwhile I was trying to fix it over the command line from the keyboard. But all of it was in vain—shortly before 10:00 I nervously dialed the technician's number.

As the "official" observer of the 2.3-m telescope, I regarded it as my duty to report the problem and look for a solution as quickly as possible. After some lengthy instructions by the tired but patient telescope specialist, we in fact managed to reboot the computer. After entering the whole lot of passwords and special VAX commands that I had never heard of before, the obedient telescope computer soon began to hum and buzz again as if nothing had ever happened. The remote diagnosis had worked and we could continue with our observations without problems. It did take us the whole night to recover from this shock, though, and we hoped nothing like that would ever happen again. Indeed, I would never hear from anyone of the telescope computer having to be completely rebooted by an observer.

8.7 Tested by Fire

In Australia it is common practice to burn back the undergrowth, tall grass, shrubs, and bushes, in a controlled fire at the beginning of the summer. This prevents the buildup of too much loose and dry material, which would help spread a bushfire very rapidly.

It is thus completely normal to find burned or blackened tree trunks on a walk or hike. They serve as reminders of past bushfires. Unlike European trees, Australian eucalyptus trees are little affected by fire. The living tissue of these sometimes gigantic trees is in the interior of the trunk, not in the outer layers. Particularly during summertime these varieties smell intensely like essential oils. If singed on the outside, they simply shed their dead layers and continue to grow. For this reason most eucalyptus trees are often surrounded by a pile of loose bark. Owing to this special survival mechanism, this type of tree has survived the bushfires that have repeatedly ravaged the country's interior for thousands of years.

Controlled back burning is also regularly done at Siding Spring Observatory because it is located on a mountain in the middle of the huge Warrumbungle National Park. In the event of a bushfire one would be trapped, which would be catastrophic, not even considering any of the valuable telescopes. To avoid such problems from the beginning, some observatories have their own fire brigades, and many employees are trained as volunteer firefighters to help evacuate people in an emergency and protect the telescopes from the flames.

In preparation for the upcoming dry summer, some back burning was being done in the national park below the observatory. It went on when I had my longest observing run of 12 nights in December 2003. I could smell the fire all day long but at sunset the small flames further down the hill were hardly visible. It was uncertain, in what direction the fire would move during the night. Then came the evening, and one of the small fires got out of control, probably whipped by a slight breeze. It slowly crept up the mountain until I could see it clearly burning in some distance from my telescope.

Soon I was observing not only stars but also the fire. In nervous suspense I waited a few hours to see whether anything would happen. By midnight, the fire had already gotten relatively close. To prevent anything from happening to the telescopes in summertime, some safety firebreaks had already been cleared around all of the observatory's telescopes. Luckily, the fire moved around my telescope's firebreak. That meant that at least my telescope at the far side of the observatory grounds was not in danger. Nevertheless, the situation did make me nervous. The fire was headed toward something else. It was moving in the direction of the wooden power pole about 50 yards away from my telescope. Many of the electric cables belonging to the telescope were suspended from that pole, and the sewer pipe ran directly by the foot of it. Problems with the power cable as well as with the sewage would not help with my observations. I had to do something.

By that point, inch-sized pieces of ash were already flying through the air, accompanied by thick black smoke. Initially, I had still been able to observe, but now, later in the night, I had to close the dome to prevent ash from flying into the building or falling onto the mirror. The wind had picked up and the increasingly intense smoke soon enveloped the whole mountain. The small unoccupied neighboring telescope was equipped with a fire alarm. When the alarm sounded it could be heard over the entire mountain. I had already called the observatory fire brigade before the alarm had started, to ask whether anything needed to be done. Meanwhile the fire was assiduously burning its way toward the power pole. After some further consideration, it was finally decided that the situation was serious enough to have the observatory fire truck come and put out the fire. In the meantime I had the small fire extinguisher from the control room ready, just in case. I was fully prepared to defend the power pole myself. Those cables were, after all, what linked my telescope with the rest of the world.

Shortly afterward, though, I realized that this plan was, of course, pretty naive. Against a formidable bushfire, you would not get very far with a small fire extinguisher designed for kitchen emergencies. Indeed, the fire truck slowly rolled up the curvy and steep road. Several fire hoses were immediately connected to the water line and the order went out, "Water! Ready!" More volunteers had arrived, and in the end we also pitched in to put out the flames that were threatening the power pole. At 2 a.m., everything was finally over.

Sadly, I could not continue to observe afterward—neither that night nor during the two nights that followed. The smoke kept lingering in the air for quite some time and thus also above the telescope. Those tiny smoke particles very strongly scatter the (short wavelength) blue light from the stars I would have liked to observe. Consequently, no blue stellar photons were reaching the detector of my instrument anymore. When the air began to clear up after a few days, everything was back to normal again. Except for the power pole, that is. It is still blackened to this day.

The moral of this tale is that you are not observing just stars when you go observing; you are also responsible for making sure that nothing happens to the observatory and telescopes at the same time. It is of no help to anyone if the power pole succumbs to a fire or if something happens to the telescope. You do have to stay calm and call the fire brigade instead. Indeed, thanks to extensive, long-term back burning and a careful watch, the observatory survived a brutal fire storm in January 2014 with minimal damage to the telescopes. However, the lodge that housed the observers fell victim to the flames.

Australian winters, too, can present problems for observations. A colleague of mine was once observing at the 2.3-m telescope. One afternoon it started to snow, which is rather unusual in Australia. Although just an inch of snow fell, the telescope had to be kept closed for safety reasons. He was writing me frantic emails about tiny snowflakes covering his telescope. Telescopes in other regions, however, are regularly snowed in, such as on the 2,600-m Mt. Hopkins in Arizona and those on the 4,000-m Mauna Kea in Hawaii. Snow and particularly ice on the roof of the telescope building are hazardous to telescope mirrors, because water can easily drip onto the mirror while the dome is being opened.

Observing is sometimes a little bit like playing the lottery. The stakes are always high, as the goal is to record a lot of new data. But being subjected to the weather, you never know whether you will go home empty-handed and have to wait an entire year for another chance.

CHAPTER 9

THE CHEMICAL EVOLUTION OF THE EARLY UNIVERSE

Our lives on Earth are in a constant state of motion, but taking a look at the celestial night sky, the stars and their positions seem to be static and permanent. The cosmos is not stationary, though, by any means, even though the timescales on which celestial objects move are gigantic compared to terrestrial and human ones.

At the beginning of cosmic evolution, there were no stars. The Universe was completely dark. Only gradually, starting a few hundred million years after the Big Bang, the very first stars began to form—the first sources of light that lit up outer space. Over the course of the next 500 million years, more stars formed, as well as the first larger stellar systems, the predecessors of present-day galaxies.

Learning about these first generations of stars requires some detective work as they are accessible only through their remnants: the chemical signature left behind after their deaths. That abundance signature is preserved in the most metal-poor stars and is possible for us to measure. By extracting this information, significant progress has been made in our understanding of how and where all the elements were first created in the early Universe, and how their production in stars has continued until today. As we will see, this story is intimately connected to our current understanding of star and galaxy formation and cosmology. We begin with taking a closer look at the lives and deaths of these first cosmic behemoths.

9.1 The First Stars in the Universe

Our knowledge about the very first stars in the Universe is currently based exclusively on detailed computer simulations and indirect observations. This is unsatisfactory, but their lifetime of a few million years was so incredibly short compared to most other cosmic timescales that they died in enormous explosions almost immediately after their formation. This makes them inaccessible to any kind of direct observation today. It is not even possible to use space telescopes to look that far into the past, into the highest redshift Universe. However, it might be possible to observe these first explosions, if they were luminous enough. Nevertheless, computer models deliver fascinating details about our cosmic predecessors and their existence. Thus, even without any observations, we have a good understanding of the fundamental physical processes governing the formation of the first stars.

About 300 million years after the Big Bang, those first objects formed from primordial gas clouds consisting of just hydrogen, helium, and trace amounts of lithium. Those clouds were of 1 million solar masses that had collapsed under their own gravity. The main issue with star formation is having the gas be cool enough to allow it to clump together. Anyone who has pumped up a bike tire knows what happens when gas is compressed: the pump gets hot. This is exactly how a collapsing gas cloud heats up if the heat cannot be radiated away. During the collapse, the gas cloud initially heats up to over 1,000 K. For (proto)stars to form from clumps, however, the gas has to be cooler than about 200 K (see chapter 4).

At the present time, the gas from which stars form is roughly 10 K, much colder than this limiting value. In the current Universe, hot gas is cooled when the very rapidly moving gas atoms are excited by collisions. It means that part of their kinetic energy is converted into internal atomic energy. After some time, this excitation energy is emitted again in the form of photons. The heavier the element is, the more electrons an atom possesses, and the more possibilities there are for internal excitation and emission. In cooler gas, molecules can form, causing this cooling mechanism of collisional excitation and emission to operate even more efficiently.

Such low temperatures could not be reached in the early Universe. The reason is that no metals or interstellar dust, which could have provided cooling effects on the gas, were available in the primordial material comprising just hydrogen and helium and traces of lithium. The neutral hydrogen atom has only one electron, the helium atom only two, and the lithium atom only three. The options of exciting these atoms through collisions are thus very limited. The only possibility was the formation of some hydrogen molecules, H_2, from individual hydrogen atoms. Molecular hydrogen was able to gradually cool the hottest inner parts of a gas cloud to about 200 K. This cooling is the result of collisions between a hydrogen atom and a hydrogen molecule, spinning up the hydrogen molecule and subsequently releasing low-energy infrared radiation.

This drop in temperature led to a lower gas pressure within the primordial cloud, allowing gravity to increase the gas density. This whole process ceased when the newly formed gas clump reached equilibrium between the outwardly directed gas pressure and the inwardly directed gravitational force. A massive protostar could eventually form out of this clump. Due to the lack of suitable cooling mechanisms in the early Universe, only large and extremely massive gas clouds could collapse under their own gravity. Accordingly, these very first stars weighed up to 100 solar masses. Low-mass stars such as the Sun could not form from such gigantic clouds.

Due to their enormous masses and special primordial composition, these first sources of light had particularly large luminosities of 1 million solar luminosities and extremely high surface temperatures of 100,000 K. For comparison, our metal-rich Population I star, the Sun, has a surface temperature of just 5,750 K. Owing to such immense luminosities, Population III stars were limited to life spans that were very short compared with most other astronomical timescales.

The core regions of the first stars were even 100 million K hot. This corresponds to almost 10 times the Sun's core temperature. The radiation emanating from these blazing giants was mostly energetic ultraviolet light, which began to heat up the neutral hydrogen and helium gas in the stars' surroundings and to ionize the atoms there. The existence of ionized gas changed the conditions of star formation in the Universe in a very dramatic manner.

Sophisticated cosmological simulations of early star formation modeling these processes have shown that the first stars were, on average, several tens to 100 times heavier than the Sun. Typical masses may have been 10 to 50 solar masses. Some of them were probably even substantially more massive. On the other hand, it might also have been possible for stars with less than 1 solar mass to emerge, although the details the formation mechanism of such low-mass objects is still being explored.

How many higher-mass and lower-mass stars formed as part of that first generation of stars remains unknown. The various cooling mechanisms in the gas determine in large part the final stellar masses, but this fact alone is insufficient for delivering the desired answers. However, the relative distribution of stellar masses is enormously important. It is one of the most important questions of modern astrophysics, since the mass distribution plays an important role in many different aspects of astrophysics.

Nevertheless, the fact that no primordial low-mass stars have yet been found is an interesting finding. The massive "residents" of the early Universe differ fundamentally from the low-mass stars populating the current Universe whose mass distribution is strikingly different. In accordance with observations, the following rule of thumb applies: the lower the stellar mass, the more frequently stars like it can exist in the Universe. Hence, stars of less than one solar mass by far dominate the Universe these days. Stars with 10 or more solar masses are consequently very rare these days. In the case of stars of more than 100 solar masses, it is unclear whether it is possible for such heavyweights to still form at all. In the past few years several gigantic supernova explosions have been observed that indicate the occasional existence of such giants even today.

The nuclear fusion processes in the interiors of the first stars were less efficient without any metals present. For example, the CNO cycle could not operate. Such a star had to be hotter and hence more compact than a metal-rich star of equal mass: a first star with 100 solar masses had a radius of just 5 solar radii. Only if a star is hot enough can the gas pressure and radiation pressure build up enough for the star to avoid the gravitational collapse due to its own mass. All these effects caused the stellar fuel to be exhausted enormously fast. Stellar evolution with its various burning stages proceeded in record time. Within a few million years those massive and metal-free behemoths exploded as gigantic core-collapse supernovae to enrich the interstellar medium in heavy elements for the first time. With the emergence of the first metals, the Universe was not primordial anymore and the conditions for the formation of all subsequent stellar generations had thus been altered forever.

Various simulations of such supernova explosions have shown that the inner portions of the outer atmospheres of stars of about 24 to 140 solar masses are not completely ejected into space. As a result, this material falls back toward the center. Since a black hole has formed there during the explosion, the gas falls directly into the black hole and, consequently, is eliminated from the cosmic cycle of matter and thus also from chemical evolution.

Stars with masses between 140 and 260 solar masses blast apart in supposedly even more energetic explosions known as pair-instability supernovae. The central temperature in these stars is so hot that highly energetic photons can spontaneously turn into electrons and positrons. The resulting loss of pressure, together with the enormous mass and temperature of the star, causes the last burning stages of such a star to proceed particularly rapidly. The outcome is not a core collapse but uncontrolled nuclear fusion which ends in a massive stellar explosion. This event is so extreme that not even a compact remnant, such as a black hole, can form. The star is completely disrupted. This theoretically predicted kind of supernova is supposed to cause an enormous metal enrichment of the interstellar medium in the vicinity of the explosion. A consequence of this specific explosion mechanism is that no elements heavier than zinc can be produced. Furthermore, lighter elements (e.g., calcium) are synthesized in different proportions than in normal core-collapse supernova explosions of stars with about 8 to 24 solar masses.

Finally, stars with more than 260 solar masses may have also existed. These supermassive behemoths must have been so heavy that instead of exploding, the entire star collapsed into a black hole and did not release any metals into its surrounding medium. For this reason, these objects, if they ever did exist, did not directly contribute anything to the chemical enrichment of the Universe.

Since it is uncertain what the overall mass distribution was among the first stars, it is very difficult to estimate the amounts of metals that were synthesized. It is also unclear what their proportions might have been and whether these elements did in fact reach the interstellar medium rather than vanishing inside the black holes. These questions are subject of current research. Elaborate models of element nucleosynthesis in core-collapse supernovae of Population III stars so far have yielded only rough answers. The processes involved are immensely complex and are a challenge not just for our own understanding of nucleosynthesis and supernova explosions but also in terms of available super-computers with which these models need to be calculated. The details of the explosion mechanism are very difficult to model and have to be approximated. Comparisons between the element abundances of supernova explosions with those of the most metal-poor stars have nonetheless already yielded many important results. Some examples will be presented in section 9.3.

The strong ultraviolet radiation from the first stars was responsible for the primordial gas becoming partially ionized. This led to the formation of molecular HD, which is composed of one hydrogen atom and one deuterium atom. HD can cool gas down farther to about 50 to 100 K. Owing to this additional cooling, there was probably a second generation of potentially metal-free stars. They likely had markedly smaller masses compared to the first generation. For the first time, stars of "only" 10 solar masses began to light up the Universe. Low-mass stars of less than one solar mass were not yet able to form, though. If the stars of the very first generation did not already do so, then members of this second generation synthesized copious amounts of metals that were deposited into the interstellar medium from their numerous core-collapse supernovae. Certainly by then was the Universe "polluted" with heavy elements once and for all. There was no turning back.

From the existence of old, metal-poor stars of less than one solar mass, we deduce that low-mass stars formed very soon after these first two generations. It thus appears that there was a transitional phase in the early Universe—from the extremely massive and therefore short-lived first stars to long-lived low-mass stars. But how did this transition occur?

The cooling of the gas cloud to below 200 K is once again of central importance. Theoretical calculations have shown that the metals carbon and oxygen are particularly well suited to cool gas. Composed only of hydrogen and helium, stars from the first generation, but possibly also some of the second generation, synthesized large amounts of carbon and oxygen during their advanced evolutionary stages. Stellar winds carried these elements from the stellar surface into the primordial medium even before any supernova explosions. The subsequent explosions of those stars further enriched the interstellar medium with carbon and oxygen. This had fundamental consequences.

In what is referred to as fine-structure line cooling, atoms excite each other onto a higher energy level after mutual collisions. An atom's "fine structure" describes when one atom has two very closely spaced energy levels due to quantum mechanical and relativistic effects. When atoms return to their ground states, they release energy in the form of a photon that can leave the gas cloud. The presence of these many fine-structure energy levels makes the gas particularly efficient at losing more and more energy, causing its temperature to drop rapidly as long as a minimum amount of carbon and oxygen are available. With carbon and oxygen, temperatures far below 200 K can be reached. This, in turn, produces regions with particularly high densities in the gas cloud. Eventually this leads to the formation of stars weighing substantially less than one solar mass.

Extremely metal-poor stars can be used to test this concept of fine-structure cooling. In all likelihood, these are stars are part of the earliest generations. If carbon and oxygen did in fact initiate this transition, then the most metal-poor stars ought to reflect it in their element abundances. Specifically, the carbon and oxygen abundances of the metal-poor stars from the early Universe should either correspond to that minimum critical metallicity or be higher. Lower abundances would not be permitted in this picture, meaning that the gas had not been cooled down sufficiently to permit the formation of those stars.

The concept of fine-structure line cooling seems largely be supported by stellar abundances. The most metal-poor stars do indeed have carbon and oxygen abundances either agreeing with the theoretically predicted minimum value or exceeding it. Moreover, the idea of fine-structure line cooling offers a tantalizing explanation for the nature of many other metal-poor stars. More than 10 years ago, astronomers noticed that almost a quarter of all metal-poor stars having less than 1/100th of the solar iron abundance ([Fe/H] < −2.0) are enhanced in carbon and thus contain at least 10 times as much carbon than iron. Five of the six stars with the lowest iron abundances have much higher carbon abundances, ranging from 40 up to 80,000 times more than iron. The exact causes of these element signatures are still mostly unexplained. They do indicate, though, that carbon played an important role for early star formation. Thus, the theory of fine-structure line cooling offers the most comprehensive interpretation of the existence of carbon-enhanced metal-poor stars to date.

The exception is an ultra-metal-poor star whose carbon and oxygen abundances are below the critical value. This indicates that probably not all stars were formed in exactly the same way from their various gas clouds. There is indeed another possibility besides fine-structure line cooling. Interstellar dust grains can also decrease the temperature of primordial gas. Dust consists primarily of carbon and silicon atoms. But these elements also had to be produced in the first stars in order to later turn into dust grains in the shock wave of the supernova. Dust cooling works only within highly condensed clumps of gas. There, it can provide an extra cooling kick for the formation of stars with less than one solar mass. The critical amount of dust in the gas is lower than that of carbon and oxygen. Therefore the existence of extremely metal-poor stars with very low carbon and oxygen abundances can be explained by this theory.

The most metal-poor stars preserve the chemical fingerprints of the first stars in the Universe in their outer atmosphere. Valuable information can thus be empirically gathered about the existence and characteristics of these first stars and their supernova explosions. This research is one of the central goals of stellar archaeology, as it affords astronomers a unique opportunity to investigate the major chemical and physical conditions of the earliest phases of star formation. Such details cannot be obtained in any other way.

9.2 The Family of Metal-Poor Stars

Detailed abundance patterns of many metal-poor stars need to be obtained to answer specific scientific questions about the origin of the elements, the involved nucleosynthesis processes, and chemical evolution. These abundance patterns consist of the ratios between the amounts of various elements determined by an abundance analysis.

Let us take a closer look at the main groups of metal-poor stars with their characteristic abundance patterns. Whereas the majority of metal-poor stars have abundance patterns typical of halo stars, about 10% present somewhat unusual patterns. These exceptions tell us about many details of the evolution of the first stars and their supernova explosions. For instance, the masses and explosion energies of the first stars can be narrowed down and estimates can be obtained about how the newly synthesized elements are mixed into the interstellar medium.

Each of these groups tells its own story and teaches us different and new details about the earliest phases of element synthesis and the sites at which they took place. Only then can the findings be compared with results of chemical evolution models obtained from various simulations of the structure and evolution of our Galaxy.

Ordinary Metal-Poor Stars

Ordinary, normal metal-poor stars form the largest group, at about 90% of all metal-poor stars. The metals in these objects make up an abundance pattern very similar to that of the Sun. The absolute abundances of the individual elements are much lower than those of the Sun, at the amount corresponding to their stellar metallicity. An important feature of these ordinary metal-poor halo stars is a characteristic enrichment of α-elements (silicon, magnesium, titanium, calcium) compared to iron at [α/Fe] ~ 0.4. It distinguishes a halo star from other stars, such as those of the Galactic disk.

This group of stars best describes the overall trends of chemical evolution in the Milky Way since their abundances reflect the main nucleosynthetic processes and their relevant contributions over long periods of time.

Carbon-Rich Stars

Carbon abundances can be measured in almost all metal-poor stars. Carbon production occurs during helium burning, independent of the production of all elements except nitrogen and oxygen.

The absorption of carbon, as well as of nitrogen and oxygen, can be detected in spectra of metal-poor stars mostly in the form of hydrides. These are the molecules CH, NH, and OH. Instead of presenting individual absorption lines, molecules produce absorption in a series of closely spaced and overlapping lines. Extended absorption bands are formed that can sometimes measure more than 10 Å. One example is the so-called G band of CH at ~4,300 Å. It can clearly be seen in the spectra of a number of stars shown in Figure 9.1. If the star is particularly rich

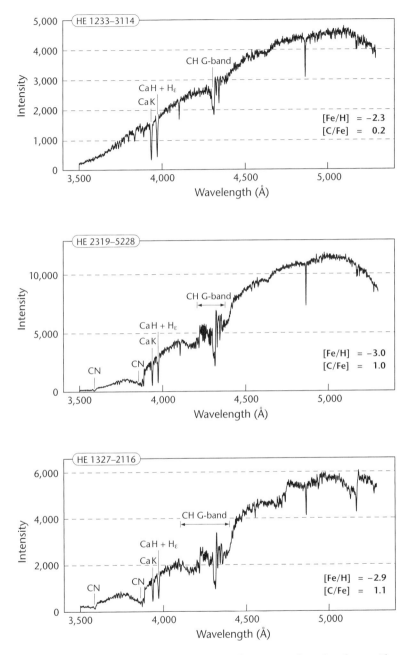

Figure 9.1. Medium-resolution spectra of stars with various carbon abundances. These spectra were taken with the 2.3 m telescope at the Siding Spring Observatory. The y-axes give the photon counts of the each spectrum. (*Source*: Peter Palm; reproduction of spectra from Frebel, "Abundance Analysis of Bright Metal-Poor Stars from the Hamburg/ESO Survey," PhD thesis, Australian National University, 2006)

in carbon, sometimes features of C_2 can be found at 5,200 Å, or of CN around 3,800 Å. Only in stars of higher metallicities is atomic carbon detectible around ~9,070 Å.

What becomes quickly apparent when working with metal-poor stars is that carbon overabundances appear in all subgroups and in all possible combinations of element patterns. About 30% of stars with ratios of [Fe/H] < –3 have 10 times more carbon than iron, hence [C/Fe] > 1. This is substantially higher than what most ordinary metal-poor Population II halo stars exhibit. In addition, countless stars have lower carbon overabundances of [C/Fe] = 0.5 to 1.0. Furthermore, the fraction of these carbon-rich stars increases with decreasing metallicity: The chance that one of the most metal-poor star is carbon-rich is thus rather high. Five of the six known most metal-poor stars with the lowest iron abundances are extremely carbon-rich.

One important question remains, though. Where does the carbon come from? In principle there are two possibilities. Either it was added to the gas cloud before the metal-poor stars formed, or else the star received the carbon at a later point in time through mass transfer from a companion star in a binary-star system. As described in chapter 5, the latter idea explains the formation of carbon-rich s-process stars. Enrichment of the native gas cloud is probably the best explanation for the group of carbon-rich metal-poor stars whose element-abundance patterns correspond to an entirely regular halo pattern for all elements with the exception of carbon.

Consequently, a lot of carbon must have been produced in Population III stars in the early Universe. The notably high frequency of carbon-rich stars among the most metal-poor stars suggests that carbon played a special role in the early Universe. The cooling effects by this element certainly contributed toward the formation of the first low mass stars in the Universe. Although the details are not yet well enough understood, carbon-rich stars permit us to study many aspects of the enrichment processes of the interstellar medium and, more generally, the carbon nucleosynthesis in massive stars of the earliest generations.

The details of the origin of the carbon overabundances in stars with [Fe/H] < –3.0 remain a hot topic of current research. In 2005 I attended a conference in which the role of carbon in the early Universe was the sole topic to be discussed for an entire week.

Stars with Special [α/Fe] Overabundances

Some stars exhibit abnormally high magnesium and silicon abundances that far exceed the normal halo star values of [α/Fe] = 0.4. This behavior mostly occurs in combination with a carbon overabundance. This fact helps to better understand the production of carbon in relation to other elements as well as the production of individual α-elements relative to one another.

Stars with Overabundances of Neutron-Capture Elements

A number of metal-poor stars with [Fe/H] < −2.0 present enormous overabundances of neutron-capture elements produced either by the r-process or the s-process, as have been described in some detail in chapter 5. In addition, some stars have neutron-capture elements produced by both the r- and s-processes. In those cases it is particularly difficult to disentangle the enrichment events that predated the formation of these stars.

Occasionally, these stars are also rich in carbon. Carbon enhancement is easily understood in connection with the s-process. When s-process elements are dredged up to the surface of a red giant star, large quantities of carbon are simultaneously moved up from the inner layers. Surface material is later transferred to the companion star. If carbon appears in connection with an r-enrichment, however, the origin of the carbon remains unclear and unexplained. The most plausible solution is that the carbon originated from a star of the previous generation but not necessarily from the progenitor supernova in which the r-process had taken place—otherwise r-process stars without carbon overabundances would be inexplicable.

Stars with Large Lead Abundances

One subgroup of s-process stars has characteristically large abundances of lead. In metal-poor stars, the s-process runs through to its very end product, lead, producing particularly large amounts of this element,

100 times more compared to the iron abundance (see also chapter 5). These lead-rich stars are in a close binary-star system and the lead was synthesized in the s-process in a somewhat more massive companion star, not the observed star itself. At a later point in time, the s-process-rich surface material of the more massive, further evolved companion has been partially transferred onto the other, now observable lower mass star.

These examples of star groups with overabundant elements illustrate the chemical diversity of the early Universe and the interplay between the many processes of nucleosynthesis. Often we know of only a few members of each group, but with time more such unusual stars will surely be found along with discoveries of new chemical groups. Ultimately, all stars assist in solving the gigantic puzzle of how the nucleosynthesis of the chemical elements is reflected in the observations of metal-poor stars. This task will likely occupy astronomers for quite a while. Hopefully someday all the abundance patterns will be precisely and uniquely associated with the nucleosynthetic processes and astrophysical sites of the corresponding elements.

9.3 The Most Iron-Poor Stars

Stars with the lowest metallicities showcase the greatest variety of unusual element abundances. Two of the most iron-poor stars are the best examples of this. But first we should consider why these objects are referred to as the most iron-poor stars and not the most metal-poor stars. Let us take a closer look at HE 0107–5240 and HE 1327–2326.

HE 0107–5240 is a red giant with an iron abundance of [Fe/H] = –5.2, which corresponds to only 1/150,000th of the solar iron abundance. HE 1327–2326 has just left the main sequence but its evolutionary stage is still close to the turn-off point. It has [Fe/H] = –5.4, just 1/250,000th of the Sun's iron abundance. Thus, there are more than 10 billion hydrogen atoms for every iron atom in the atmosphere of HE 1327–2326. All told, this star contains a total of 100 times *less* iron than the iron core in the Earth's interior. This is quite little, in light of the fact that this star is about 300,000 times heavier than Earth.

If, as is usually done, one assumes that a star's iron abundance equals its metallicity, both these stars ought to be the by far most metal-poor ones. What we have learned from these discoveries, however, is that most elements in these stars do not emulate iron. As a matter of fact, both these iron-poor stars exhibit extreme ratios of carbon, nitrogen and oxygen to iron, as well as relatively large amounts of sodium, magnesium, calcium, and titanium compared to iron. Adding the abundances of all these elements, the two stars suddenly become the exception to the rule: They are substantially more metal-rich than a metallicity suggested by just the iron abundance. The general rule "iron abundance = metallicity" breaks down.

What is quite exciting in this context is the latest discovery (in 2014) of a star, SM 0313–6708, so deficient in iron that no spectral lines of iron could be detected in its high-quality spectrum. Only an upper limit on the iron abundance was obtained. It is a record low [Fe/H] < –7.3. It corresponds to less than a ten-millionth of the solar iron abundance. For the first time, we have been faced with the dilemma that there was nothing measurable in the spectrum—at least in terms of iron. Molecular CH features could be detected and they showed, once again, that the most iron-poor stars are carbon rich. It has the highest ever measured [C/Fe] ratio of at least 80,000. Oxygen is even more extreme being at least 250,000 times more abundant than iron (other detected elements are lithium, magnesium, and calcium). Clearly, the iron abundance in this star is a poor measure for its overall amount of metals. These kinds of striking discrepancies have only occurred in stars with [Fe/H] < –5.0. With more discoveries of similarly iron-poor stars being sought, it will be interesting to learn whether or not their iron abundances would reflect the total metallicity. Either way, the results will be of far-reaching importance for our understanding of the formation of the first low-mass stars in the Universe.

The enormous overabundances of carbon, nitrogen, and oxygen are worth examining a bit more. Consider the values for HE 1327–2326 when it has 2,500 times more carbon than iron and 5,600 times more nitrogen than iron in its atmosphere. Oxygen also exists and is as much as 630 times more abundant. The origin of these types of overabundances in all the most iron-poor stars is still not completely understood. But models of stellar evolution and supernova nucleosynthesis offer quite plausible explanations for the production of CNO elements in the first stars and thus the observed element signature of HE 1327–2326.

What is the situation with the other elements? When a star evolves to become a red giant, its lithium is destroyed in the deeper layers of the star due to convection operating (see also section 9.4). It is thus no longer detectable in giants such as HE 0107–5240. The warm giant SM 0313–6708, however, does show a weak lithium line, in accordance with expectations based on its evolutionary stage. Since HE 1327–2326 is still in the pre-red-giant stage, lithium should be detectable. With its low iron abundance, HE 1327–2326 ought to be an ideal candidate for measuring the value of primordial lithium abundance. Surprisingly, the lithium doublet line was not detected in its spectrum. Instead of an unusual overabundance of a particular element, there is now an unexpected major deficiency found. Even after more than a decade since its discovery, there is still no satisfactory explanation for the lack of lithium in the star, although the idea of lithium having been processed through the first stars is promising. On the other hand, lithium is not depleted beyond expectations in SM 0313–6708 which has an even lower iron abundance. This puzzle will remain a challenging research topic in the years to come.

Along the same vein, it is interesting that, contrary to all expectations, another star has been found in the meantime: SDSS J102915+172927, at [Fe/H]= –4.8 has no detectable lithium either. One can only speculate that lithium is possibly destroyed in the stellar interior prior to the giant branch stages, by processes very specific to stars most deficient in iron.

Finally, HE 1327–2326 surprisingly contains the neutron-capture element strontium, in an amount 15 times greater than iron. It remains completely unclear where such a large amount of strontium could come from. A special type of supernova would likely be required, perhaps one only occurring in the early Universe. In HE 0107–5240, by contrast, no strontium could be measured.

Apart from the three stars with [Fe/H] < –5.0, we now also know of three other stars with metallicities around [Fe/H] = –4.8. What do their abundance patterns look like? Do they also have such highly individual element patterns suggesting that chemical evolution was still in its infancy and had not yet assumed its regular course? The answer is yes and also no. One of these stars, HE 0557–4840, closely resembles HE 1327–2326 and HE 0107–5240, at least in its carbon abundance.

Otherwise, though, it is less conspicuous and most of the other element abundance ratios are the same as those of halo stars with $[Fe/H] > -4.0$. The other star, SDSS J102915+172927 (containing no lithium, as described above), exhibits abundance ratios just like any other ordinary metal-poor star. At first glance it looks boring, but the conclusion to draw from this is an interesting one: a transition to more regular halo-star abundance ratios appears to take place between $[Fe/H] = -5.0$ and -4.5, perhaps from the second to subsequent generations of stars.

Most stars with higher metallicities have the regular halo pattern, although special cases obviously do exist, as has been described in section 9.2. Compared to that, among stars with the lowest iron abundance there are no normal cases, only exceptions. Put differently, the most metal-poor stars teach us that the very early Universe was still inhomogeneous and likely not yet well mixed. This is why HE 0107–5240 and HE 1327–2326 became desirable test objects immediately after their discoveries. They are extremely well suited to testing models about stellar and Galactic evolution while also offering new input to a variety of other questions.

The mere existence of these stars immediately poses the most important question in stellar archaeology: Can these exceptional abundance patterns be attributed to the chemical enrichment of their native gas cloud by just a single Population III first star? Various models for supernova explosions of Population III stars have been developed to explain the origin of these unusual patterns. The nucleosynthetic products thought to be responsible for the observed abundance patterns can thus be estimated. The overarching goal is to simultaneously reproduce the very different values for iron and carbon observed in the two stars. It is a big challenge to generate such large amounts of carbon while producing only very little iron at the same time.

Nonetheless, a breakthrough came with the scenario about an exploding Population III star of 25 solar masses whose newly synthesized elements are not all ejected far enough into the interstellar medium during the supernova explosion. Consequently, some portions of the gas, particularly layers deep within the star such as iron, fall back onto the collapsing stellar core. They are swallowed immediately by the nascent black hole. This way, only tiny amounts of iron enrich the surrounding gas, whereas other elements, such as carbon, can

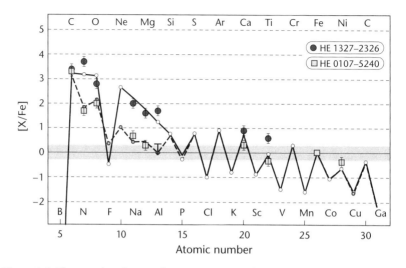

Figure 9.2. Element abundances of HE 1327–2326 and HE 0107–5240 (circles and squares). The thick dashed lines indicate the best possible fits for abundances calculated with a nucleosynthesis model of a core-collapse supernova with 25 solar masses in which not all the gas was ejected into the interstellar medium. The horizontal solid line at [X/Fe] = 0 indicates the solar abundance ratios for comparison. The arrows indicate upper limits of the observed element abundances. (*Source*: Peter Palm; data from Nomoto et al. *Nuclear Physics* A 777 (2006): 424. With kind permission of Ken'ichi Nomoto)

occur in much larger quantities. A comparison between the stellar abundances of HE 0107–5240 and HE 1327–2326 and predictions for such supernova yields of a Population III star is shown in Figure 9.2. The good agreement suggests that the gas clouds from which HE 0107–5240 and HE 1327–2326 formed had in fact each been enriched by just one such first star.

Particularly for the case of HE 1327–2326, another idea to explain the carbon overabundance considers rapidly rotating Population III stars with 60 solar masses. Due to their enormous mass loss prior to the supernova explosion, large amounts of carbon as well as nitrogen and oxygen can be released into the interstellar medium.

The method of comparing theoretically predicted with observed element abundances is the only way to learn about chemical evolution and to reconstruct its earliest phases. It is quite a challenge to find these rare objects that may in fact be the survivors of the second generation of stars in the Universe. But the reward for our understanding of the early Universe and the first generation of stars is unique and priceless.

Finally, the abundance patterns of stars with metallicities of [Fe/H] = −4.0 and above with their typical halo abundances *cannot* be explained by a single progenitor star. In order to reproduce those abundance patterns, averaged values for the nucleosynthesis products of several supernovae appear to be required. The variations in the individual nucleosynthesis products average out to a uniform mix only once a larger number of supernovae have contributed their elements. The chemical evolution of the Universe has been evolving ever since.

9.4 The Cosmic Chemical Evolution

The Big Bang left behind a universe made of only hydrogen, helium, and traces of lithium—everything else was somehow made in stars born since then. The resulting chemical evolution of the Universe continues to this day. Stars of different metallicities in the Milky Way and in various dwarf galaxies help us reconstruct the countless processes involved in this evolution. The most metal-poor stars tell us about the earliest eras in the cosmos, while metal-rich stars tell us about the later times. Hence, the metallicity [Fe/H] can be taken as a rough measure of the time elapsed since the Big Bang. In order to understand element production and the observed abundance trends, we will take a closer look at the processes of nucleosynthesis responsible for the evolution of all the different elements.

In the following figures, the horizontal axis indicates stellar metallicity. It allows us to easily follow the trends and behavior of the various elements over time. The stellar iron abundance not only serves as a proxy for the total metal abundance of a star but also relays information about the enrichment timescales of the gas from which new stars have successively formed over the course of many billions of years. By interpreting the abundance patterns of metal-poor halo stars in combination with the patterns of dwarf galaxy stars, comprehensive insights can be obtained regarding the details of this complex chemical evolution. This is true not only for individual galaxies but also for the Universe as a whole.

In the following sections, the evolution of the most important elements and element groups over the first 4 to 5 billion years of the Universe is summarized.

Helium

Helium is the second most abundant element in the Universe. It is not a metal in the astronomical sense. Unfortunately, no spectroscopic measurements of helium in metal-poor stars exist that could record the evolution of helium over time. For reasons based on atomic physics details, He absorption lines appear in spectra only if their stars are hotter than ~7,000 K. Furthermore, such hot stars already have undergone countless mixing processes that altered their surface composition. This precludes measurement of helium in the gas cloud from which a star had formed. In cooler stars, only chromospheric He lines are measurable. Overall, determining a He abundance is rather complicated for various reasons and often entirely impossible.

Nevertheless, the evolution of this element is interesting, as helium is constantly being synthesized in stars and being expelled into space by planetary nebulae and supernova explosions. The best possibility of measuring the primordial helium abundance is based on spectra of thin interstellar clouds at high redshift. The range of helium abundances by mass spans 23.2% to 25.8%, with 24.9% being the currently best value.

Lithium

Lithium is an important element, but it is notoriously difficult to assess. Although it can be measured in metal-poor stars, its evolution in the early Universe is unclear and controversial. Lithium is very easily converted into other elements by proton capture in a star's hotter interior regions. The overall stellar production rate of lithium is thus rather meager. Accordingly, when it comes to lithium there is no real chemical *evolution*. All lithium in the Universe, even terrestrial lithium, instead originates from the Big Bang. On Earth, South America has the greatest exploitable lithium deposits in the form of rocks and argillaceous earths. Furthermore, lithium is also present in seawater. Without the Big Bang lithium, the lightweight yet powerful lithium-based batteries probably would not exist. Lithium would also be sorely missed in various industrial and pharmaceutical applications.

Lithium detected in metal-poor stars obviously traces the earliest lithium production events. We can thus examine this element in the hope of learning about the primordial lithium abundance. During the 1980s it was found that there is always a specific amount of lithium in metal-poor stars. Yet, it is 2.5 times lower than the prediction of the primordial lithium value based on data from the WMAP satellite that measured the anisotropy of the cosmic background radiation, in combination with predictions about standard Big Bang nucleosynthesis.

Since the primordial lithium abundance has been very precisely determined, it is generally assumed that the lithium measurements of metal-poor stars do not directly reflect the primordial lithium value after all. Why metal-poor stars have so much lower abundance levels remains largely unclear, though. Any future explanations of this long-standing discrepancy will hopefully lead to a better understanding of how early stars and galaxies have influenced the amount of lithium in the Universe.

CARBON, NITROGEN, AND OXYGEN

Carbon, nitrogen, and oxygen are dispersed into space not only during stellar evolution through stellar winds but also during the supernova explosions of massive stars. As has been presented in chapter 3, carbon atoms are initially synthesized by the 3α-process in red giants during advanced stages of stellar evolution. Oxygen is produced in parallel with carbon by the α-process, whereby another α-particle is captured by the carbon nucleus. Consequently, oxygen can be regarded as an α-element. Abundance analyses of metal-poor stars have shown that oxygen does in fact act like other α-elements as it evolves over time. In contradistinction, nitrogen is produced in the carbon(-nitrogen-oxygen) cycle.

A star's rapid rotation causes mixing of the outer layers that enhances the production of carbon and nitrogen. Stellar mass plays an additional role. It is thus not surprising that massive rotating Population III stars very likely produced large amounts of carbon already during the earliest times. In later eras, the less massive stars of 3 to 8 solar masses became the main producers. Present in large numbers, these stars are currently supplying the interstellar medium with carbon and other elements by their stellar winds that are particularly powerful during the asymptotic giant branch stage.

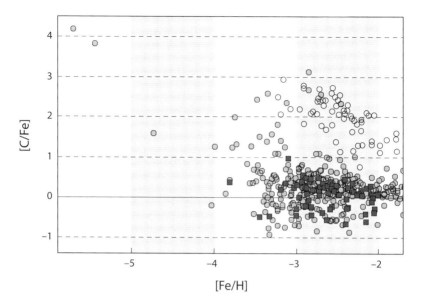

Figure 9.3. Carbon abundances [C/Fe] of stars with different [Fe/H] metallicities (filled circles). Open circles denote carbon-rich s-process stars as well as stars with both s- and r-process enrichments. R-process stars usually do not differ from the other stars in their carbon abundances. They are represented as squares. The continuous line indicates the solar [C/Fe] ratio for comparison. (*Source*: Peter Palm; data from Frebel *Astronomische Nachrichten* 331 (2010): 474–488)

The evolution of carbon can be seen in Figure 9.3. Stellar metallicity is increasing along the horizontal axis. Along the vertical axis astronomers show a given element's ratio to iron, such as [C/Fe]. It makes it easy to spot whether a star deviates from an element pattern typical of halo stars which have a carbon-to-iron ratio of about zero, [C/Fe] ~ 0, but with a spread ranging from [C/Fe] = −0.6 to +0.6. In the figure only metal-poor stars with [Fe/H] < −1.7 are shown and the continuous line indicates the solar [C/Fe] ratio as a reference. The high [C/Fe] abundances of the most metal-poor stars are clearly recognizable as well as the relatively large number of stars below [Fe/H] < −3.0 with somewhat higher carbon abundances. S-process stars as well as stars enriched by the s- and r-processes are marked by different symbols. They received their carbon from their binary companions, and thus they clearly show a different level of carbon abundances than other stars, setting them apart in the figure. R-process stars are not specifically marked because, with one exception, they all have carbon abundances similar to ordinary halo stars.

Even if we knew nothing about any nucleosynthetic processes, looking at Figure 9.3 would already reveal one important fact: carbon was produced in many different ways and in several types of stars in the early Universe. Otherwise such a variety of [C/Fe] ratios would not be found among metal-poor stars.

A similar picture is beginning to emerge for some dwarf galaxies. Even though up to now only about a dozen stars at [Fe/H] < −3.0 have been observed in the various dwarfs, two strongly carbon-rich stars at [Fe/H] = −3.2 and −3.7 have already been found. Evidently, extremely metal-poor, carbon-rich stars occur not just in the Galactic halo. One can thus conclude that these stars probably represent a more general signature of the early phases of chemical evolution in the Universe.

α-ELEMENTS

α-elements, including magnesium, silicon, calcium, and titanium, are composed of a multiple of helium nuclei. They are synthesized during various burning stages in massive stars. Model computations of nucleosynthesis have confirmed that α-elements and iron are produced in a very distinct ratio of [α/Fe] ~ 0.4. Core-collapse supernova explosions later eject these elements out into space. α-elements can easily be measured in high-resolution spectra of any star, since these elements have relatively strong absorption lines even in metal-poor stars.

Abundance analyses have already shown long ago that the majority of metal-poor stars have enhanced α-element abundance ratios compared to the Sun. Figure 9.4 illustrates this behavior. Stars with [Fe/H] < −1.5 have [α/Fe] values typical of core-collapse nucleosynthesis, which yield [α/Fe] ~ 0.4. In contrast, some stars with higher metallicities have lower [α/Fe] ratios, whereas stars with solar metallicities finally exhibit solar α-element abundances, that is, [α/Fe] = 0.

How can this behavior be explained? The cosmic evolution of the [α/Fe] ratio is one of the best examples of chemical evolution and the interplay between various nucleosynthesis processes and sites as well as different enrichment timescales.

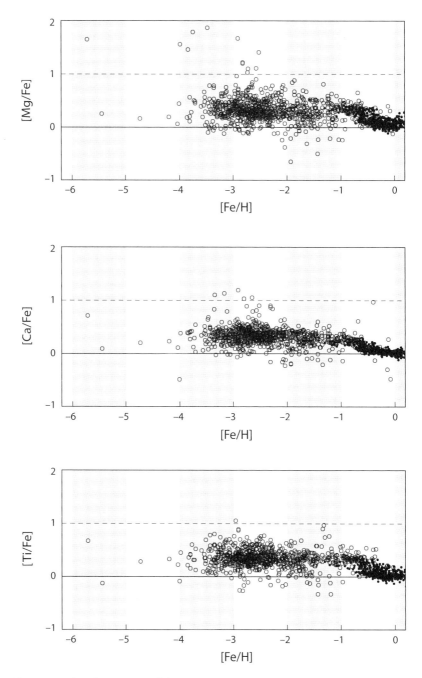

Figure 9.4. Abundance ratios of the α-elements [Mg/Fe] (magnesium), [Ca/Fe] (calcium) and [Ti/Fe] (titanium) for halo stars of different [Fe/H] metallicities (circles). More metal-rich stars in the Galactic disk are shown as smaller dark dots for comparison. At about [Fe/H] ∼ –1.0, all three element ratios of [α /Fe] begin to drop from a value of ∼ 0.4 to the solar value (continuous line). This change is due to the onset of Type Ia supernova explosions, as these explosions produce iron and no α-elements. At [Fe/H] ∼ 0.0, chemical evolution has reached the solar ratio of [α/Fe]. (*Source:* Peter Palm; data from Frebel *Astronomische Nachrichten* 331 (2010): 474–488)

The chemical evolution of the early Universe was driven exclusively by short-lived massive stars and their core-collapse supernovae explosions. Metal-poor stars with their [α/Fe] ~ 0.4 values precisely reflect this early period. Owing to their longer lifetimes, stars of lower mass were only halfway through their evolution at that time. The first lower mass stars only became white dwarfs roughly one billion years later. If these white dwarfs belonged to a binary-star system and received a mass transfer from the companion, they eventually exploded as a Type Ia supernova. Type Ia explosions changed the overall course of chemical evolution because there are many more low-mass stars in the Universe than massive stars that will explode this way. These supernovae mainly generate carbon, oxygen and elements of the iron group, but no α-elements. This means that the onset of white-dwarf explosions caused the cosmic production of iron to increase significantly. It is precisely this radical change that we can see in the abundances of α-to-iron ratios in stars of different metallicities. α-elements continued to be produced by massive stars, but from this moment forward iron was being produced not only by massive supernovae but also by the exploding low-mass white dwarfs.

The rise in the cosmic iron abundance reduced the values of [α/Fe] in the gas clouds from which new generations of stars would be born. The resulting transition at [Fe/H] ~ −1.5 from the "early," high [α/Fe] ratios in metal-poor stars to lower ratios in the somewhat more metal-rich stars at later times reflects this evolution. The younger metal-rich stars with solar metallicities finally reached the solar [α/Fe] value.

Naturally, some exceptions do also exist. Individual metal-poor stars with extremely high magnesium abundances, for example, are being discovered. Such signature is probably caused by unusual kinds of supernovae that enriched the gas cloud in a special way before the birth of the metal-poor star. Other exceptions include metal-poor stars exhibiting lower α-to-iron abundance ratios than a typical halo star. A few such stars can be encountered in the Milky Way, but most of them are found in dwarf galaxies.

Dwarf galaxies undergo chemical evolution just as the Milky Way does. Since these small dwarfs have less gas available for star formation, their entire evolution proceeds more slowly. Even so, the local white dwarfs in binary systems also start to explode after about one billion years because stellar evolution proceeds independently of the evolution of a galaxy in which a star is located. Therefore, after that billion years, dwarf galaxies had a total metallicity lower than that of the Milky Way since its chemical evolution did not advance as far. The consequence of the longer timescales of enrichment in dwarf galaxies is that the characteristic transition from higher α-to-iron values to lower ratios occurs at metallicities that are lower than the Milky Way's value of [Fe/H] = −1.5.

The precise transition value depends on the galaxy and is often difficult if not impossible to determine. Dwarf galaxy stars with values between [Fe/H] = −2.0 and −2.5 are good candidates for such measurements. Recent studies have shown that dwarf galaxy stars at [Fe/H] ~ −3.0 and even lower metallicities also show higher [α/Fe] ratios typical of halo stars. It thus appears that chemical evolution proceeds in a similar way during the early phases of galaxy formation, namely by means of core-collapse supernovae. Later on it continues at different rates, as small galaxies need longer than larger ones for large-scale element production.

Iron-Group Elements

The elements of the iron group, namely, scandium, vanadium, chromium, manganese, iron, cobalt, nickel, copper, and zinc with atomic numbers from 23 to 30, are synthesized in massive stars. This happens in the last burning stages of stellar evolution, for instance during silicon burning and also during the supernova explosions when different kinds of nucleosynthesis processes take place in the region surrounding the shock wave. The abundances of scandium, chromium, and cobalt are shown in Figure 9.5. The continuous line indicates the solar abundance ratio for reference. The cobalt-to-iron abundances in the most metal-poor stars at [Fe/H] ~ −3.5 are, on average, higher ([Co/Fe] ~ +0.5). As the metallicity increases, [Co/Fe] decreases slowly and reaches the solar value at about [Fe/H] ~ −2.0. Zinc shows the same behavior. Therefore, in the early Universe elements were synthesized more abundantly compared to iron. Chromium and manganese exhibit the opposite behavior. The most metal-poor stars have the lowest [Cr/Fe] and [Mn/Fe] abundances. The solar value is only reached at [Fe/H] ~ −1.0. This means that compared to iron, less chromium and manganese were produced in the early Universe. Scandium and nickel exhibit a different behavior still.

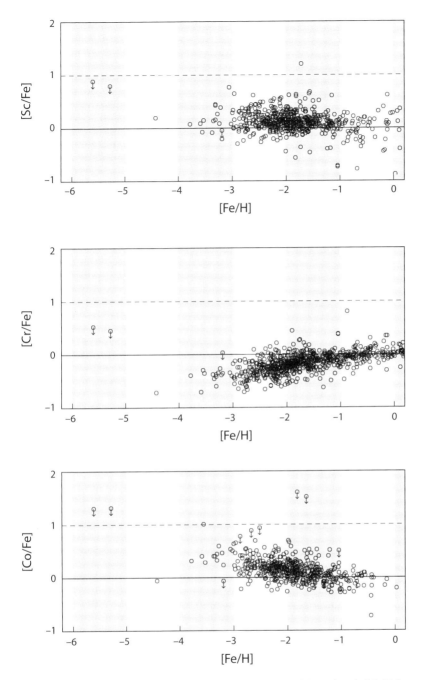

Figure 9.5. Abundance ratios of iron-group elements [Sc/Fe] (scandium), [Cr/Fe] (chromium), and [Co/Fe] (cobalt) for halo stars of different [Fe/H] metallicities (circles). Arrows indicate upper limits of the observed element abundances. The evolution of iron-group elements differs very much from one another and also from that of the α-elements. The continuous line indicates the solar abundance ratios for comparison. (*Source*: Peter Palm; data from Frebel *Astronomische Nachrichten* 331 (2010): 474–488)

The [Sc/Fe] and [Ni/Fe] ratios remain the same at all metallicities at the solar value. Metal-poor stars in dwarf galaxies display the same behavior and thus do not differ in any way from the halo stars of the Milky Way.

Despite having a different evolution, the abundance trends of iron-group elements are well-defined and with few outliers. The elements that share the same behavior were likely produced by the same kind of nucleosynthetic process. Furthermore, scandium and nickel have in all likelihood been synthesized in the early Universe by exactly the same processes that still operate at the present time. Nevertheless, the vastly different behaviors between iron-peak elements are somewhat unexpected and are still not understood. Attempts have already been made to explain these differences, for instance, with models of altered supernova properties, such as the explosive energy—but as yet without much success.

Neutron-Capture Elements

Elements heavier than zinc occur in the Universe only in trace amounts compared to the lighter elements: They are about one million times rarer than iron, for example. As has been described at great length in chapter 5, smaller amounts of these heavy elements are built up incrementally in two different processes through a successive capture of neutrons. The r-process likely operates in core-collapse supernova explosions, whereas the s-process works in evolved giant stars of ~3 to 8 solar masses.

Irrespective of their quantities, all elements play their role in the chemical evolution of a galaxy. In a unique manner, each element reflects the delicate interplay between all astrophysical processes and sites of nucleosynthesis contributing toward the formation of the elements. The story of the evolution of neutron-capture elements as conveyed by metal-poor stars is accordingly complex. A simple interpretation is not possible here.

Figure 9.6 depicts the evolution of barium. Different symbols distinguish s-process stars from those stars with s- and r-process enrichments. Barium is a main s-process element—accordingly, these stars have huge barium overabundances up to 1,000 times greater than iron. As the nucleosynthetic origin of neutron-capture elements for these stars is clearly known, it is not surprising that they are separated from the other stars in Figure 9.6. The same applies to r-process stars, which

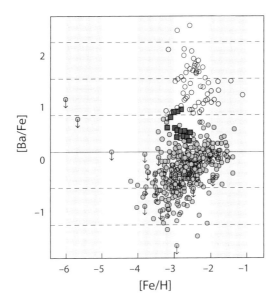

Figure 9.6. Abundance ratios of the neutron-capture element barium ([Ba/Fe]) of halo stars with different [Fe/H] metallicities (filled circles). The arrows indicate upper limits of the observed element abundances. The open circles signify s-process stars as well as stars with both s- and r-process enrichments. Squares label r-process stars. Both groups have very large barium abundances, by virtue of their classification as stars with large amounts of neutron-capture elements. The evolution of neutron-capture elements is entirely different from that of lighter elements: there is an enormous spread of more than a factor of 100,000 between the abundances of stars with the lowest and highest [Ba/Fe] values. The continuous line indicates the solar abundance ratios for comparison. (*Source*: Peter Palm; data from Frebel *Astronomische Nachrichten* 331 (2010): 474–488)

are marked by squares. Barium is generated by the r-process, too, but in smaller amounts.

Where does barium come from in ordinary halo stars? From the s-process or the r-process? The answer to this question is rather complex. One thing is clear, though: the range of barium abundances reflects an immense diversity of possible processes similar to the s- and r-processes and their respective production rates. Accordingly, it is easy to assume that over time all these processes contributed to the general enrichment of the interstellar medium. All the ordinary halo stars formed from gas enriched by many different processes prior to their formation. The metal-poor stars thus reflect the average level of the chemical evolution of neutron-capture elements in the Universe and not any individual nucleosynthesis events.

As Figure 9.6 shows, stars with [Fe/H] ~ −3.0 have very low barium abundances compared to iron and to the Sun. With increasing metallicity, the barium abundances also rise. Nevertheless, this trend is not particularly well defined, compared to the trends of α-elements and iron-group elements (see Figures 9.4 and 9.5); at each metallicity, there are stars with a large range of barium abundances. It will likely take a few more years of intense research about the various nucleosynthesis processes to eventually explain all these observational results in a satisfactory manner.

Carbon, Hydrogen, Oxygen, Phosphorus, Sulfur, and Nitrogen—The Elements of Life

Almost all the elements in the periodic table, with the exception of hydrogen and lithium, are generated in stars and then "recycled" by being integrated into cosmic objects of the next generation. This describes the overall chemical evolution in the Universe. As has become apparent from the research into metal-poor stars, the amount of the elements present in the early Universe did not suffice, by a long shot, to yield a planet such as Earth, upon which life could eventually develop. Earth, composed of iron, oxygen, silicon, magnesium, and other elements, could actually form only once the Universe had sufficient amounts of these particular elements. However, we know that the right amount was produced sometime within the first ~9 billion years after the Big Bang. Otherwise the Sun and the Solar System could not have formed from the presolar nebula 4.6 billion years ago.

Chemical evolution is one of the most complex processes in the Universe and even the current best models and simulations can only coarsely describe exactly how it proceeded in detail. It is still unclear at what point the Universe was "mature" enough, from the chemical point of view, for the emergence of the first planets. In addition, planet formation itself is a fairly complicated and complex process. Moreover, all planets of the Solar System have different chemical compositions, showing that there is no simple "recipe" for planet formation.

Current extensive searches for planets around other stars are being followed with excitement and suspense by the public at large. Stars resembling the Sun need to be observed over long periods of time to reveal whether they host planets. In their selection, extremely metal-rich stars are mainly sought whose metallicities are higher than that of the Sun. A large amount of metals present in the birth gas cloud increases the chance of finding a star being orbited by one or even several planets, one of which could perhaps host life. In the past few years, the Kepler space telescope has already discovered hundreds of planets, although none among them is very similar to Earth. Finding the right one seems, however, only a matter of time. Hence, even for the search for Earthlike planets, the analysis of the chemical composition of the host stars is of great importance.

How did those elements crucially involved in the emergence of life as we know it evolve? The most important elements in this context are carbon, hydrogen, oxygen, phosphorus, sulfur, and nitrogen. The human body is mainly made of these elements. Due to the large percentage of water in the body, oxygen takes the lion's share of the body's mass, at 61%. Carbon comes in second at 23%, then hydrogen at 10%, and finally nitrogen at 3%. The remaining 3% constitutes many different trace elements. The chemical characteristics of carbon make it combine easily with other elements to form molecules, particularly with hydrogen, nitrogen, and oxygen. This way complex molecules needed in all organisms, such as proteins, nucleic acids, carbohydrates, and fats, can be formed. Consequently, these three elements are extremely important for life, and for maintaining a breathable planetary atmosphere containing enough oxygen to sustain life.

Phosphorus and sulfur also play an important part in the human body. Both elements fulfill fundamental tasks in every cell by enabling the formation of important molecules. Without phosphorus there would be no DNA and RNA molecules. The metabolic exchange of energy in a cell does not operate without phosphorus. So what is known about the cosmic history of these two elements? First of all, it is difficult to detect phosphorus and sulfur in metal-poor stars. Some of their absorption lines are hidden in the near-infrared region among many lines of H_2O (i.e., water) and other molecules. These are generated in the terrestrial atmosphere, not by the star itself. Such contamination complicates precise line measurements in the spectrum. Moreover, the generally weak lines are not detectable in extremely metal-poor stars. However, some phosphorus lines are present in the UV spectral range for which data can be obtained only with space telescopes. New results have recently shown that phosphorus can be measured from the UV lines even in extremely metal-poor stars, enabling a complete reconstruction of the cosmic evolution of this element.

Phosphorus can be made from silicon atoms by neutron-capture. It occurs mainly in later burning stages during the evolution of massive stars. Newly synthesized phosphorus is ejected into space by the subsequent supernova explosion. As for sulfur, it technically is an α-element and is thus built up, just like magnesium or titanium, by the capture of helium nuclei. Accordingly, sulfur is likewise produced in core-collapse supernova explosions and then dispersed into the Universe. Consequently, phosphorus and sulfur have been present in the Universe since the earliest times. As they cannot form in low-mass stars or in other burning or evolutionary stages, their production rates have changed little with time, if at all.

Let us finish with an amusing little story. The main task when working with metal-poor stars is to determine the metallicity of each object. Can we do the same for the human body, too? A colleague once posed this question, asking his audience to raise their hands to indicate whether they thought that human beings are metal-poor or metal-rich. Both sides received 50% of the vote. Any body's metallicity is, of course, defined by the ratio of iron to hydrogen. Our iron is in our blood, and hydrogen is a component of water, which makes up more than half the mass of our bodies. Therefore, we are metal-poor compared to the Sun—we have a metallicity of [Fe/H] = −0.5, which is three times less iron than hydrogen compared to the solar ratio.

CHAPTER 10

FINDING THE OLDEST STARS

Thus far we have investigated stars and their evolution over time, examined the Milky Way with its many inhabitants, introduced the methods of spectral analysis, and considered the beginning of star formation in the early Universe. We have also become acquainted with the diverse groups of metal-poor stars and learned how these objects help us reconstruct the time soon after the Big Bang and how they are used to extract details about the chemical evolution of our Galaxy.

What is still missing is the eventful discovery story of one of the most iron-poor stars, HE 1327–2326. After all, I was directly involved in pushing the scientific boundary of the field of stellar archaeology with new exciting results. Finding old stars, analyzing old stars, interpreting old stars—these tasks continue to fascinate me as much as ever. Hardly anything can lure me away from pursuing the oldest, most metal-poor stars. Until early 2014, HE 1327–2326 held the record for the most iron-poor star known. But as it happens in science and in life, records are meant to be broken. The discovery of the latest record holder by our team is described in section 10.4.

10.1 Pursuing Metal-Poor Stars

Compared to the Andromeda galaxy, the Milky Way is home to four to five times fewer stars. However, the many hundred billion stars in our own Galaxy still represent a substantial collection. In any case, confronted by these large numbers, astronomers have developed various stellar classification schemes to categorize them. There are numerous types of stars after all. Sorting them into groups and classes has aided enormously in understanding and interpreting the wealth of observed stellar spectra. Annie Jump Cannon classified thousands of stars based solely on surface temperature. But other fundamental differences also exist among stars.

Around 1944, the German astronomer Walter Baade divided all stars into two groups: Types I and II. The main distinction was the strengths of the absorption lines of metals compared to the strengths of the hydrogen lines in a given star's spectrum. Another decade had to pass, however, before the basic characteristics of Type II stars with their weaker absorption lines would be understood. Only then did it become possible to develop a physical explanation for this new kind of classification.

Until about the middle of the last century, astronomers worked under the assumption that all stars had the exact same chemical composition as the Sun. It was thought that the only differences were surface temperatures, which led to different strengths of the absorption lines in the observed spectra. In the 1940s, some halo stars were found with metal absorption lines that, strangely enough, were substantially weaker than the lines in the solar spectrum. This could not be explained by assuming all stars were chemically identical. Conjectures were made that these stars might have substantially less hydrogen than helium compared to normal stars, or else that they might have unusual outer atmospheric layers.

In 1951, the American astronomers Joseph Chamberlain and Lawrence Aller proposed something revolutionary. They concluded—as "one possibly undesirable factor" of the interpretation of their stellar spectra—that the stars in question had "abnormally small amounts of calcium and iron." They had measured only about 1/20th of the solar calcium and iron values in two of their stars.

This unusual suggestion confused many contemporary astronomers. Let us recall what the state of the art in physics and astronomy was at that time. This was the period predating the 1957 B^{2}FH article on nucleosynthesis, although it was already known that nuclear fusion occurred inside stars. Even though the Big Bang theory had already been established by 1950, people still widely assumed that all elements had been formed from primordial gas shortly after the Big Bang. Nobody was thinking of any chemical evolution in the Universe yet, let alone that it could be studied by means of stars with different metallicities.

Some of my colleagues still enjoy recounting the story that Chamberlain and Aller had actually found metallicities of only 1/100th of the solar iron value for their stars. But such a result seemed so outrageous to the referee of their paper that they altered the surface temperature enough to increase their measured metal deficiency from 1/100th up to between 1/10th and 1/30th. It worked because a star's temperature influences the spectral line strength of an element, although they were not able to reach a solar metallicity this way. As Chamberlain himself later admitted, the fudged higher value was in fact published in 1951. Interestingly enough, the originally measured metallicity of 1/100th comes very close to the value accepted today. HD 140283 is one of the bright stars that Chamberlain and Aller had observed. Many subsequent analyses throughout the last half century have clearly demonstrated that this subgiant star is indeed metal-poor and contains only about 1/300th of the solar value: Its metallicity currently stands at [Fe/H] ~ −2.5. Today, HD 140283 is indisputably a classical metal-poor halo star and continues to be used frequently in chemical analyses as a reference star. I have also observed and analyzed this star several times.

Much new research carried out in the following two decades finally demonstrated that a wide range of stars with different metallicities and element-abundance patterns exists. The diversity was soon attributed to various stages of the chemical evolution of the Milky Way. The foundation was laid for researching the early Universe this way. Today, this work is called "stellar archaeology." In retrospect, the stellar spectroscopy work of the 1950s actually transformed the entire worldview at that time. The belief in a chemically homogeneous Universe was now transformed into one experiencing a chemical evolution that would drive the evolution of galaxies and that of the Universe as a whole. Astronomy these days is inconceivable without considering the effects of a star's, and even of an entire galaxy's, metallicity.

Baade's star groups are still being used today but referred to as Population I and Population II. They roughly reflect the chemical evolution of the Milky Way: Population I is the substantially larger group, as it describes young and metal-rich stars primarily found in the Galactic disk. Older, metal-poorer stars from the halo with their weaker spectral lines form Population II. Table 10.1 summarizes the populations of stars.

Table 10.1. Definitions of Stars with Different Metallicities

Type	Definition
Population III	First generation of (metal-free) stars
Population II	Old (halo) stars with low metal abundances
Population I	Young, metal-rich (disk) stars, e.g., the Sun

During the 1980s astronomers first proposed that a third population also existed. It was supposed to consist of the very first stars of the early Universe. Half a dozen stars with [Fe/H] ~ −3 were known at the time, but no objects with even lower metallicities. Thus, it was thought that any of these so-called Population III stars would have metallicities less than [Fe/H] = −3. After all, an iron abundance ratio of less than a thousandth of the Sun's seemed tiny enough to be an appropriate composition for one of the very first stars.

Although simple models of the chemical evolution of the Milky Way predicted the existence of such first stars, initial efforts to find them remained unsuccessful. With some desperation, the American astronomer Howard Bond titled an article in 1981 "Where Is Population III?" Bond had been attempting in vain to track down those particularly metal-poor stars in a first systematic survey. But he made no discoveries. The initial conclusion was that long-lived stars of less than one solar mass could hardly form from primordial gas in the early Universe. If they ever existed, they must be extremely rare, otherwise Bond would have found some of them.

Indeed, the various predictions by Milky Way models available at that time had significantly overestimated the number of stars most deficient in metals. In addition, Bond, using the telescopes of his day, had been able to observe only relatively bright stars. One rule of thumb to remember is this: the fainter a star, the farther out it is in the halo. Experience shows that the probability of finding a star of low metallicity rises with its distance. Furthermore, simply surveying a larger volume of fainter and fainter stars increases the number of metal-poor stars that one is able to identify. Consequently, Bond's chances of finding a star with [Fe/H] < −3 were extremely low.

Around 1980, the star CD −38° 245 was rather coincidentally found by the Australian astronomers Michael Bessell and John Norris. They later called the discovery one of "informed serendipity." They determined the star's metallicity to be [Fe/H] = −4.5, corresponding to less than a ten-thousandth of the solar iron abundance. This value was so low that it seemed hardly possible for even more iron-poor stars to be discovered. Accordingly, in 1984, it was proposed that a Population III star had at last been found. Evidently, Bond's endeavors had not been misguided. This discovery soon led to the term "Population III" being equated with extremely metal-poor stars containing such miniscule amounts of metals.

Today, "Population III" is again used exclusively when referring to the very first stars. A lot of the theoretical research and corresponding cosmological simulations concern the formation and evolution of the very first, extremely massive objects that are metal-free stars. A clean definition based on a physical concept could thus be introduced: members of the Population III are metal-free stars consisting only of hydrogen, helium, and traces of lithium. It immediately follows that the most metal-poor stars are the most extreme examples of Population II stars, and do not actually belong to Population III. Chapter 9 is devoted to the Population III first stars, though.

A few years ago, I had the pleasure of meeting Howard Bond myself. It was with great joy that we discussed the search for metal-poor stars and his pioneering research. Considering that the scientific article about his search had been submitted to the *Astrophysical Journal* in the year of my birth, it was wonderful to hear his take about the beginnings of my field of work. He seemed to be quite pleased that his research has been successfully continued and that stars with record-breaking low metallicities have been found since then. Although from a technical point of view these discoveries cannot be designated as Population III stars owing to their containing trace amounts of metals, his quest from more than 30 years ago ultimately did meet with success.

Using increasingly comprehensive sky surveys, the hunt for more metal-poor stars in the halo has intensified since the 1980s. The coveted stars with metallicities down to [Fe/H] ∼ −3.8 were quickly found, but it took almost another 20 years for the German astronomer Norbert Christlieb and his colleagues using the Hamburg/ESO Survey to break the record set by CD −38° 245. In 2001, the star HE 0107–5240, having just 1/150,000th of the solar iron abundance ([Fe/H] = −5.2), was a sensational find. Its discovery turned out to be a breakthrough in this field of research: At last it was clear that such chemically extremely primitive metal-poor halo stars are in fact observable in our own Galaxy. In other words, a witness of the earliest chemical enrichments in the Universe had finally been found.

Three years later, in 2004, our team discovered the star HE 1327–2326 under my leadership. At [Fe/H] = −5.4, its iron abundance lies at just about 1/250,000th of the solar iron value. This record has only very recently been surpassed, as a series of projects are currently under way with the goal of finding more of these extraordinary stars. The team around Australian astronomer Stefan Keller found this new star that has such a low iron abundance that only an upper limit on the iron abundance, [Fe/H] < −7.3, could be derived. Three additional stars have also been discovered since 2007, all with iron abundances of [Fe/H] ~ −4.8, at last beating the longtime record holder CD −38° 245.

In a certain sense, CD −38° 245 has by now been relegated to a comparison star: I own a collection of spectra that I always take along on observing runs. They serve as an immediate visual comparison against which each newly observed star is compared. The goal is to quickly determine if the star exhibits unusually weak lines. If they are similarly weak to those in CD −38° 245, I get excited and even a bit nervous. Most of the time, though, the lines are considerably stronger, which means that the star is considerably more metal-rich and hence not as interesting. From such simple comparisons it is easy to determine whether it is worthwhile to spend more precious telescope time on a candidate star.

Figure 10.1 summarizes the discovery history of the most metal-poor stars since Chamberlain and Aller in 1951. One star with a substantially lower iron abundance has been found about every 20 years, although progress has sped up with our latest star discovery in early 2014. The suspense of how this story will continue is unabated as it is unclear when (and whether) any new record breakers will be found. The next generation of giant telescopes are sure to play a part in this. The steady trend toward ever lower metallicities is reflected, to a certain degree, in the mirror sizes of the telescopes available to astronomers at their time: from small telescopes with 1- to 2-m mirrors around 1980, to ones with mirrors measuring about 4 m and then to 6 to 10 m since about the mid-1990s.

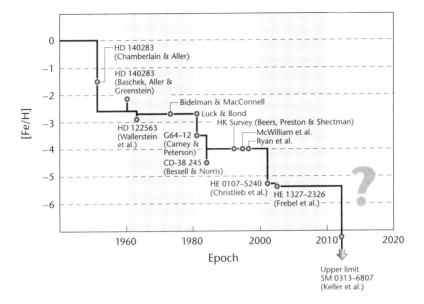

Figure 10.1. The iron abundances of the most metal-poor stars known at the time. Dots indicate the abundances originally reported by the authors, whereas the interconnecting horizontal lines indicate the currently accepted values. (*Source*: Peter Palm; reproduction of data from Frebel and Norris, "Metal-Poor Stars and the Chemical Evolution of the Universe," in *Planets, Stars and Stellar Systems*, New York: Springer, 2013)

How many metal-poor stars have been found and observed in this way? Compared to the few hundred billions of stars in the Milky Way, we are of course not talking about large quantities. Nonetheless, the following numbers (as of mid 2014) reflect the success story of stellar archaeology whose aim is to find these needles in the Galactic haystack. Hundreds of stars with [Fe/H] < –3.0 have been discovered to date, but only about 200 of them, namely the brighter ones, have been observed and analyzed with high-resolution spectroscopy. Stars with [Fe/H] < –3.5 are significantly rarer, with the consequence that detailed analyses exist for all the 45 known examples. Just 8 stars are known with [Fe/H] < –4.5, 3 of which have [Fe/H] < –5.0 and 1 that has [Fe/H] < –7.3. The most interesting stars by far are those with metallicities of [Fe/H] < –3.5. They provide us with the best insight into the early evolutionary history of the early Universe.

All these discoveries have enabled the field of stellar archaeology to thrive, particularly over the past 10 years. Yet, many unresolved questions remain. One of them is what the lowest observable metallicity of a star could be, besides the very first metal-free stars. With the recent discovery of a star with [Fe/H] < −7.3, we are close to answering this question. Additional discoveries of similarly iron-poor stars with actual iron abundance measurements will provide more definitive answers, though.

10.2 Bright Metal-Poor Stars

The stars of the Hamburg/ESO Survey had been divided into fainter and brighter stars. The spectra of the latter suffered from saturation effects of the photographic plates used in the survey. The erstwhile record holder HE 0107–5240 was one of the fainter stars and for that reason had already been discovered by my German colleague. The sample of brighter stars had been entrusted to me "to take a look. Maybe something interesting is in there." The goal of my PhD thesis was to process this already pre-selected stellar sample, in order to identify and later analyze any truly metal-poor stars. Although this task does not sound that difficult or lengthy at all, this short statement summarizes about three and a half years of my life. In the following paragraphs, I will explain why this search was so time-consuming and outline that the devil often lies in the details.

At the beginning of a new project it helps to have a general idea or a rough feeling for the anticipated outcome. I find the prospect of pursuing a particular goal to be a great motivation when embarking on any new project. Only then can I begin my work with the necessary enthusiasm and perseverance. After all, when carrying out novel scientific research you cannot simply check the solution at the back of a textbook or ask your teacher.

Scientific research actually requires you to be very creative. You have to constantly imagine how certain processes operate in the Universe. In most cases, even observations can give you only an instantaneous snapshot of a given situation. This requires constantly making new hypotheses that then have to be tested. Creativity is usually exclusively associated with the arts, but science would still be in its infancy if researchers had not thought of countless novel and creative ways of how to examine and understand the world and the cosmos. Hence, every experiment is a work of art because it is the realization of a new idea.

Accordingly, many projects in astronomy go back to an initial motivation or idea. A certain amount of risk is always involved, but it is the only way to acquire new knowledge—even when the actual project does not produce the anticipated result. However, many studies are designed to minimize risk. Important scientific results are obtained this way, which provide fruitful grounds for further work and ideas. Examples of such projects include, for instance, studies of environmental effects on the evolution of cosmic objects or establishing trends and correlations. These kinds of investigations for the most part ensure a solid result. In contrast, when it comes to discovering new, rare objects, there is not really any guarantee of success, or only insofar as some new objects will surely be found but not necessarily the ones you had hoped to find. Hence, the plan for my dissertation also included another project with already available data. The purpose was to ensure that, in any event, I would be able to obtain a significant scientific result. But my main task was to search for and find metal-deficient stars.

The preselection of metal-poor candidates had been made by my colleague Norbert Christlieb by means of computer programs, as part of the overall processing of the Hamburg/ESO Survey data. I was highly motivated and immediately began to work on the low-resolution survey spectra. The prospect of finding ancient stars from the early Universe was extremely thrilling to me. The first major task was to assess every single spectrum of the preselected sample on the computer screen. The aim was to sort all of the 5,500 objects into different classes. This task was quite tedious, but my enthusiasm was boundless. This was rather fortunate because the total inspection lasted two long weeks.

My sample contained the spectra of particularly bright stars. The brightness of many of the objects caused the photographic plates that had been used for the survey to become completely saturated (see Figure 10.2). The resulting effects caused a loss of information, and hence, reduced the quality of the spectra. It was not even clear whether the sample was at all usable for the search for metal-poor stars. My task was to provide a definite answer to exactly this question.

The saturation effects had a particularly severe disadvantage that ultimately yielded many falsely classified candidates. The more deficient in

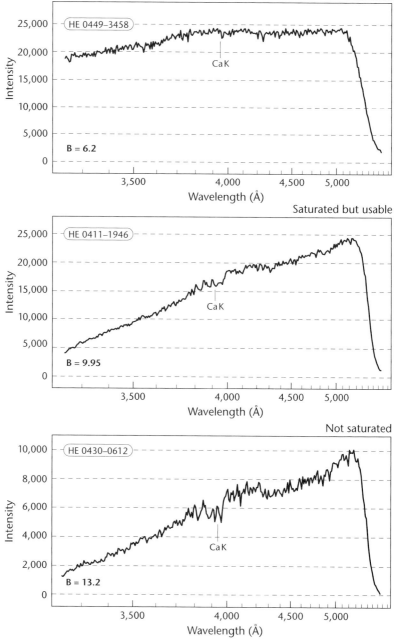

Figure 10.2. Comparison between partially saturated and "normal" survey spectra. Saturation effects occur with particularly bright stars. The uppermost spectrum is almost completely saturated because it runs almost entirely horizontally. The middle spectrum displays a saturation at the right end but the region important in the search for metal-poor stars, at about 3,900 Å, is unaffected. The bottom spectrum is not saturated. These latter two stars were classified as *mpcc*. The y-axis indicates the photon counts of each spectrum. (*Source*: Peter Palm; reproduction of spectra from Frebel, "Abundance Analysis of Bright Metal-Poor Stars from the Hamburg/ESO Survey," PhD thesis, Australian National University, 2006)

metals a star is, the weaker are its absorption lines. A saturated spectrum also, falsely, displays very weak absorption lines, as can be seen in Figure 10.2. Such spectra had been classified as metal-poor. Accordingly, my visual inspections of the entire sample concluded that 3,733 objects were not really metal-poor but particularly bright, very hot metal-rich stars with saturated spectra. A few galaxies classified as stars, as well as a variety of objects with artifacts in their spectra, were among them, too. Although the problems associated with saturation effects had been foreseen, the outcome was initially quite disappointing. Only 1,777 candidates remained in the end. All I wanted was to find as many metal-poor stars as possible, but now my sample had shrunk to a third of what I had started with.

During the inspection, I had divided those 1,777 stars into different metal-poor classes (mpc), depending on how metal-poor their spectra appeared to be: *mpcc* for stars with a relatively strong calcium K line at 3,933.6 Å, *mpcb* for stars with a weak calcium line, *mpca* for stars with no visible calcium line, and *unid* for spectra where it was unclear whether a calcium line was visible due to noise in the data. Searching for the most metal-poor stars, these last two categories were the most promising. In all cases *mpc* abbreviates "metal-poor candidate," with the alternative of "unidentified" for the ones with ambiguous calcium lines in the spectrum. In the end I found 1,426 *mpcc* stars, 248 *mpcb*, 84 *unid*, and a grand total of 9 *mpca* candidates. Not particularly many stars, but not so few either!

From scrutinizing 5,500 spectra, I quickly learned that every spectrum looks a little different and that each star does have something of its own personality. It did take a while, though, before I, new to the art of spectroscopy, had familiarized myself enough with spectra to be able to classify them quickly and with a certain amount of self-confidence. To some extent this task resembled the classification work by Annie Jump Cannon and her coworkers. The whole procedure could be imagined as resembling passport inspections at the airport. A software program first displays the spectrum on the screen. As the inspector, you then check it for various characteristics and in the end the star gets stamped as either "granted" or "denied." Ultimately, this classification was a pretty responsible duty. Stars that were denied, in other words thrown out, were eliminated once and for all and never looked at again. This meant that an erroneous classification could very well have the consequence that a potentially very interesting star was discarded because it had not been recognized as such. Yet the investment involved in reinspecting all those rejected objects in detail one more time is simply too large to make sense for such kinds of survey projects.

After my inspection of 5,500 spectra, I was glad to be finished. For a long time afterward, though, I hoped that I had not inadvertently discarded any or at least not too many interesting stars. A big question mark remained whether, out of inexperience or erroneous judgment, I had failed to select some stars. However, because I did find some quite interesting stars in my sample in the end, I can in retrospect assume that I probably did not do a terrible job with the classifications. From a statistical point of view it seems rather unlikely that my sample contained many more of the most metal-poor stars besides my discoveries.

The worry about perhaps having discarded useful metal-poor candidates illustrates two more things that I had to learn during my thesis research. On one hand, one has to always accept some losses when working with large-scale surveys, no matter how much effort is expended. The quality of the data is simply low, and quantity beats quality in this case. Consequently, any false classifications simply cannot be completely excluded. I find it a bit frustrating to this very day, but there is no changing it. On the other hand, I learned from this experience that such large research projects do ultimately bear fruit. With an exciting goal in mind and confidence in the subject and one's ability, anyone can find something new and fascinating in the end.

10.3 Mt. Stromlo Succumbs to Bushfires

Early in the month of August 2002 I arrived in Canberra from Germany for some scientific work experience at Mt. Stromlo Observatory. At that time, I could not suspect that this observatory was destined to be consumed by a bushfire and largely destroyed just five months later under extraordinary circumstances. It was during the days immediately following the fire that I finished the selection of my bright metal-poor candidates and began my work on the whole sample. For that reason, my recollections of the beginning of my search for metal-poor stars are inevitably linked with this tragic event.

Nobody who grows up in Germany, as I did, will encounter serious fires. You may hear about them or see some images on television, but people are never seriously worried each summer that their home could possibly burn down. To me, major blazes were something from apocalyptical films, in which handsome men covered in grime and soot resist the flames by the sweat of their brows and save human lives. That is, up to that day of the raging fire in Canberra.

But back to that Australian summer in 2003, when on Saturday, January 18, I was at home in O'Connor, a suburb of North Canberra. Bushfires had been burning for weeks already, since mid-December 2002, in the national parks roughly 50 km away. During the daytime the sky was occasionally rather dark. Smoke was seen on the horizon and the air often smelled like burned wood. But nobody was particularly anxious about it, as such bushfires occur frequently in Australia and are entirely normal there.

On that Saturday morning, however, I saw particularly large clouds of smoke rising that soon covered the whole sky. They were much blacker than those the days and weeks before. Hitherto, they had always been coming from the Namadgi National Park southwest of Canberra, a large part of which was already in flames by that time. With my neighbors I was keeping an eye on those smoke clouds, half out of curiosity, half out of concern as they gradually became thicker and thicker. On that hot summer morning I was not seriously worried, though, as it was apparently "just a bushfire from Namadgi Park." So I was repeatedly and reassuringly told by my Australian friends and colleagues.

At around 3:00 in the afternoon, I was still convinced that we would be having a barbeque that evening in the suburb of Duffy with some of the other students. Duffy is the westernmost suburb of Canberra and closest to the observatory. The observatory was just on the western outskirts of Canberra, in the middle of a small plantation pine forest on the low-lying Mt. Stromlo. My radio was playing in the background all day, and at around 3:30 we suddenly heard a loud siren through the radio. A calm voice repeatedly advised its listeners to immediately return home and protect themselves and their homes against bushfires. A state of emergency had been declared for Canberra. What I did not know at that moment was that Mt. Stromlo Observatory had already burned down one hour before!

The screaming sirens were sounding every 15 to 20 minutes now to inform people about the latest position of the approaching 35-km-wide wall of flame. Advisories were being given about what to do in an emergency and which parts of the city were already on high alert. They designated which areas the fire had either already entered the city or were still in great danger. Since our house in O'Connor was situated behind another hill, closer to the city center, we were in an unaffected part of town.

Around 5:00 p.m., a coworker called me from downtown. He had been living in Duffy and the police had come to evacuate the residents in his area. With the police officers breathing down his neck, he had just been able to pack the barest of essentials in a small suitcase. Then he had aimlessly driven to the center of town and did not know what to do next. I immediately invited him to stay with us in O'Connor for the time being. We soon decided to try to save more of his belongings from his house.

The sky had turned almost completely black that summer afternoon. The street lamps in the neighboring part of town had switched on but many of the traffic lights were not working. Only a few cars were still in the streets with their headlights on, creeping slowly and carefully through the darkness. Bits of ash and charred leaves were flying about in large quantities. Plate 10.A shows the state of affairs at that time. As it was very hot, almost 90 °F, and my car had no air conditioning, I attempted to briefly open the window. It did not turn out to be a useful idea, though, owing to the rather heavy and smoky air.

When we arrived in the western part of Canberra, all the major streets had already been closed off. The police advised us to turn back, but I did not want to give up so quickly. We then tried to reach Duffy through another part of town. There we were stopped by the smoke and fire itself. We found ourselves in such a thick and dense cloud of smoke that we could not see a thing around us. Some eucalyptus trees directly next to us on the sidewalk were burning, along with the grass and shrubbery underneath them, and the ashes were being blown directly at us.

We turned around and along yet another detour we at least made it to the edge of Duffy. There, we happened to meet several other Mt. Stromlo students. They were standing in the middle of the street waiting for their house to burn! Parts of the fence were already in flames, and the house next door was already gone. The sight was heartbreaking. The police had evacuated my friends from their house and would not let them go back to try to fend off the flames. It was a terrible moment, seeing them so helpless and desperate.

During such bushfires, most houses are not immediately ignited by the rapidly moving fire front but catch fire only when flying branches land on the roof or in the garden. Usually, those burning branches have the time to continue burning to then set the house on fire. I heard the next day that in the last minute my student friends were fortunately permitted to extinguish those small "spot fires" around their house, which ultimately saved it. Satellite images of the area from sometime later depicted their gray roof surrounded by a large black circle, the burned garden. The neighboring property was one large blackish-gray rectangle of cinders.

We never actually made it into Duffy that evening, despite all our efforts. In anguish, we had to believe that my coworker had lost his house and personal belongings that afternoon. It wasn't until the following morning that he found out his house had been spared. But the house next door was burned to the ground.

The big problem that day had been the strong wind. It came directly from the west at 80 to 90 km/h. At last, around 7 p.m., the wind calmed down a bit, then came from the southeast. As a result, the danger that the fires would be swept into other parts of the city diminished. Still, Canberra lost a total of 490 houses that afternoon and another 300 were damaged. Thousands of people were affected in one way or another. There were also 500 injuries and even four deaths. The worst was over. Not for me, though. Two days later, Monday morning, I heard on the radio that Mt. Stromlo Observatory had burned down! The inconceivable had happened: in just 20 minutes, a 40- to 50-m-high wall of flame had swept over Mt. Stromlo and transformed nearly the entire observatory grounds into charred rubble.

There I stood, in tears, incredulous. "My" observatory had burned down and some of my friends and acquaintances had lost everything. This fact sank in only very slowly. Feeling incredibly fortunate to be allowed to work at Mt. Stromlo to conduct professional astronomical research, I was utterly devastated. All of a sudden I, too, was affected by the bushfire and its rampage. Invaluable items, historic telescopes, data, the scientific results of years of hard work—everything had been destroyed in but a few minutes. No one had anticipated it. The place had been completely unprepared when it was hit by the fire.

For the next three weeks all astronomers were temporarily housed on campus downtown and provided with computers and Internet connections so that we could continue working. Then we were informed that for some inexplicable reason the two office buildings on the mountain had been spared by the blaze. But as I was able to witness myself the following day, five historically valuable telescopes were sooty wrecks, their mirrors lay broken on the ground and everything was covered in ash. Plate 10.B shows some of the burned-down telescopes. The Commonwealth Solar Observatory building, a heritage-listed building from 1924, had also been reduced to burned ruins. Plate 10.C compares the building before the fire, directly afterward, and its reconstruction, completed some five years later. As can be seen in Plate 10.D, bookshelves were still standing in the library but the books were nothing but ash after an incredibly hot and rapid fire passed through—a slight jog was all that was needed to make the books disintegrate into the ash they were. The machine shop had also gone up in flames. Tragically, it had housed an almost finished instrument that had cost several million dollars and was about to be shipped to the 8-m Gemini Telescope. It was now a piece of charcoal.

During the cleanup of the mountain, I began to work on my bright star sample in the provisional computer room on campus. Normal daily life resumed only after we could return to those office buildings on our charred mountain two weeks later. Everything smelled like smoke for long afterward, and it was constantly depressing and emotionally draining to see destroyed telescopes all around us. Nonetheless, all of us were glad to be back on the mountain again. The tragic situation also strengthened our small astronomy community. We helped those who had lost everything by contributing household items and moral support as we began to turn our attention to the observatory's reconstruction.

In spite of this terrible event, I suddenly felt even more a part of the Stromlo community, although I was merely an exchange student at the time. This sense of belonging motivated us to work hard and produce good science. We wanted to show everyone that we were not going to give up easily.

It soon became clear that the reconstruction would take years. The damage came to around AUD$75 million. Some eight years later the last new buildings were finally completed, and now only a few traces of the fire are visible. I enjoy visiting each time as I have many wonderful memories of my time there. The new buildings do, of course, make this small astronomy oasis look very different than in my days, up until 2006, when the rebuilding was just getting started. But no matter what it looks like, Mt. Stromlo Observatory will remain!

10.4 The Discovery of a Record-Breaking Most Iron-Poor Star

The aim of my PhD thesis work was to examine my by then diminished sample for metal-poor stars. The three-step procedure described in section 7.5 was crucial in successfully isolating the most interesting, most metal-poor star candidates.

After the initial selection, the next step was follow-up observations of all 1,777 candidates to obtain better spectra for a more precise measurement of the calcium line. A few colleagues helped me out with some of the follow-up observations to make faster progress. Whenever I had accumulated another larger batch of spectra from various different observing runs, including my own, I analyzed them in groups to determine their metallicities and select candidates that were indeed metal-poor. Throughout this work, stars kept cropping up with spectra clearly showing the object to be far too hot to be a genuine metal-poor star. They were rejected right away, of course. The spectra of other stars occasionally indicated that they were hot too, but not quite like other more typical ones. Wanting to do nothing wrong, I noted the names of these hotter "problem" stars for later discussion with my colleague Norbert Christlieb.

One of these stars had the well-sounding name HE 1327–2326. Its spectrum was quite similar to that of a hot star with intrinsically weak lines. The first analysis suggested that it had quite low metallicity. But up to that point, only stars with surface temperatures inadvertently measured as too hot had such erroneously low values. Yet I found that this spectrum somehow looked different from a typical false positive. For reasons I cannot quite recall anymore, this star remained on my list unnoticed for several months. I had been pretty swamped with other tasks, such as additional follow-up observations, to wrap up the observational part of my dissertation as quickly as possible. This list was nothing more than a little piece of paper upon which I had scribbled a few star names, along with some other random notes. Unfortunately, the original note got lost during one of my intercontinental moves, but I can still remember it clearly.

The unusual star was on that list and I was only awaiting my colleague's approval to finally discard it from the sample. Since four eyes can see more than two, I was cautiously waiting for our next meeting, which happened in May 2004. Together with three other astronomers, we had been invited to Michigan State University to work there for two weeks and discuss various projects. It was a good opportunity to consider the aforementioned problem stars with my colleague and also to report on the progress with my thesis work. On one of the first days I showed him the follow-up spectrum of HE 1327–2326, along with the associated analysis results. My colleague stared at the spectrum and gasped for air. Then came a torrent of statements, such as "Oh, wow, this is extremely interesting!!" and "We need a high-resolution spectrum right away!"

What followed happened rather quickly. First of all, I contacted my observer colleague with whom I had often observed at Siding Spring Observatory. As chance would have it, he was there and in a position to immediately supply us with a better, medium-resolution follow-up spectrum. We wanted to be absolutely sure that nothing had gone wrong with our version of the spectrum. It was raining in Australia when we first made our inquiry, but we were lucky. For about 10 minutes it was dry and clear enough to observe this bright star with the 2.3-m telescope. It was the only star that my colleague was able to observe during his entire three-night run.

We immediately concluded that the new spectrum looked exactly like the old one. Nothing had changed, but it was possible to see the very weak calcium line even more distinctly. Figure 10.3 depicts this spectrum. The new analysis indicated HE 1327–2326 to be relatively hot, and hence a main-sequence star with an iron abundance of $[Fe/H] = -4.3$.

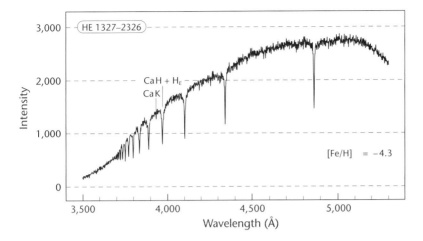

Figure 10.3. The spectrum taken with the 2.3 m telescope confirming that HE 1327–2326 is an extraordinarily metal-poor star. The tiny calcium K line can be seen between the hydrogen lines at about 3,900 Å. The y-axis indicates the photon counts of the spectrum. (*Source*: Peter Palm; reproduction of spectra from Frebel, "Abundance Analysis of Bright Metal-Poor Stars from the Hamburg/ESO Survey," PhD thesis, Australian National University, 2006)

What a moment that was! Contrary to all expectations, within a matter of two days the "garbage star" had become the most important object of all our research activities. Everything happened so fast that I was hardly able to notice the pivotal change that this star had brought to my research. At that time, there was one known star, CD −38° 245, with [Fe/H] = −4.0 (a slightly corrected value from the original analysis), and another one, HE 0107–5240, with [Fe/H] = −5.2. All other stars had higher iron abundances. Finding a star with an iron abundance in between these two stars would have been sensational. Other scientists had already started to speculate about whether any stars would ever be found between [Fe/H] = −4 and [Fe/H] = −5.

The next step was to obtain a high-resolution spectrum to verify the iron abundance as soon as possible. I cannot even talk about luck anymore: One of our Japanese colleagues happened to fly to Hawaii just a few days later to work with the 8-m Subaru Telescope and its high-resolution spectrograph. The aim of his observing run was to search for extremely metal-poor stars—a program that had been in progress for several years. The telescope time had already been granted in 2003, and so it was an extraordinarily lucky coincidence that I had just found this unbelievably good candidate right in time for these observations. The colleague immediately agreed to add the star to his program and to give it the highest priority among the observations.

What ensued resembled life in the fast lane—at least from the scientific point of view. We received the high resolution spectrum, and HE 1327–2326 was, in fact, metal-poor. Indeed, it was actually a record breaker! We had *over*estimated its iron abundance with the medium-resolution spectrum. Interstellar calcium in the gas located between us and the star was clearly discernible in the spectrum that could not have been detected from the 2.3 m telescope follow-up spectrum owing to its lower resolution. Thus, in reality, the star had much less calcium to offer than we had thought. Luckily, we could now see two tiny iron lines in the high-resolution spectrum which implied that we were also no longer dependent on a metallicity estimate based on the calcium line anymore.

In total, we could detect only four iron lines in the whole spectrum. Owing to its enormous iron deficiency, the star has almost no iron in it. The fairly warm surface temperature further reduced the strength of the lines. Nevertheless, we determined the iron abundance of HE 1327–2326 to be $[Fe/H] = -5.4$, corresponding to just 1/250,000th of solar iron abundance. It was hard to believe. I had found this new record holder for the most iron-poor star during the first year of my doctoral research work. Figure 10.4 shows the part of the high-resolution spectrum with the calcium K (and interstellar calcium) line as well as the strongest iron line.

Even though my thesis was aiming precisely at such an event, I could hardly believe it. The thesis overview that I had been required to submit when I began my research had indeed included a brief paragraph to that effect. It stated that in the event of a discovery of a star with $[Fe/H] < -5$, all the other projects should be put aside, so that I could fully concentrate on the new discovery. Exactly this exciting but rare case had occurred.

As expected, I had to do a lot of work following this sensational discovery. A detailed analysis had to be conducted and published as rapidly as possible. Another star with $[Fe/H] < -5.0$ was an extremely important

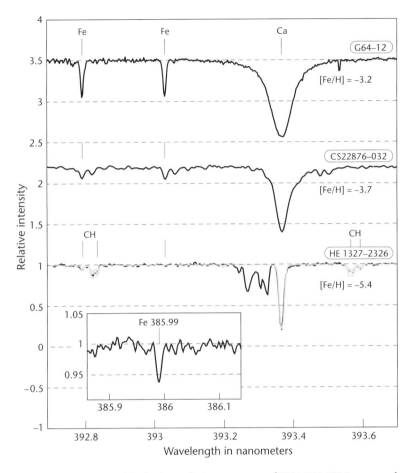

Figure 10.4. A portion of the high-resolution spectrum of HE 1327–2326 compared to that of a similar but more metal-rich star, G 64–12. Various absorption lines around the calcium K line are marked. In this region in HE 1327–2326, molecular carbon (CH) lines have appeared in lieu of iron. The gray line, which largely overlays the spectrum of HE 1327–2326, is a synthetic spectrum produced with the star's abundances. It reflects the observed spectrum very well. The non-synthesized region is interstellar calcium present in the gas between Earth and the star. The small cutout contains the spectral portion with the strongest iron line. All the spectra are normalized and some are shifted. (*Source*: Peter Palm; reproduction of spectra from Frebel et al., *Nature* 434 (2005): 871–873)

finding, showing that the discovery of HE 0107–5240 had not simply been luck. A class of stars with such minuscule amounts of iron existed, and our search methods proved once and for all that we could indeed uncover them. Finding just one star might have been chance. But discovering two stars meant that we were very well able to unlock the secrets of the early Universe with our search techniques. The details of the abundance patterns of these chemically extremely rare stars have been described in detail in section 9.3.

Late in the summer of 2004, I then spent 6 weeks in Japan together with my German and Japanese colleagues, to work on the analysis and the manuscript. The discovery was finally published in the scientific journal *Nature* in April 2005 by our team of 19 collaborators under my direction. In the meantime, three more articles have been published in astronomical journals that concern additional aspects of HE 1327–2326 and its existence, and many more present various interpretations. Hence we continue to learn much about the early Universe from these stars.

In May 2005, my colleagues and I participated in a large international conference in Paris that most scientists in our field attended. Figure 10.5 shows the group photo of the main members of our team. At the conference, I gave a talk announcing the discovery of HE 1327–2326 in front of 200 scientists from all over the world. I explained its abundance pattern and its nucleosynthetic interpretation—namely, the assumption that just one of the first stars was responsible for the observed elements and their quantities.

Until early 2013, the record set by HE 1327–2326 had not been broken. Thanks to our ongoing survey work using the SkyMapper Telescope at Siding Spring Observatory in Australia, a new extraordinary star, SM 0313-6708, was recently discovered. Selected from the survey photometry among 60 million stars, the candidate looked extremely promising after a medium-resolution spectrum had been taken with the 2.3-m telescope. There was no discernible calcium line in that spectrum, making it an exciting target for high-resolution spectroscopy with the Magellan Telescope. Indeed, soon thereafter such a high-resolution spectrum could be obtained. By lucky coincidence, we had telescope time about one month after the 2.3-m telescope observations. I was one of the first to inspect the new data. I actually panicked a little because I thought

Figure 10.5. Group photo of our international metal-poor-stars team (Australia, Japan, Germany, Britain, and United States) at a symposium of the International Astronomical Union in 2005 in Paris. (*Source*: Anna Frebel)

mistakes had been made with the observing or data processing—there were hardly any absorption lines in the spectrum. Clearly, something had to be wrong because I could not find the calcium K line. Well . . . I eventually found it after taking another very close look. The calcium line was extremely weak, much weaker than anything I had ever seen before. Two tiny magnesium lines were there, too, and plenty of CH absorption. But not much else. Iron lines were nowhere to be found despite the good data quality. After that adrenaline rush I emailed and called up some other team members to immediately discuss the news. The subsequent analysis yielded an upper limit on the iron abundance of [Fe/H] < −7.0 which meant that I had lost my record with HE 1327−2326. I was happy and sad at the same time, but more happy because our team had broken the record and not someone else. This is what I had wanted and excitement immediately took over. Since SM 0313−6708 also turned out to be immensely carbon rich, its chemical signature fit right in with the two stars with [Fe/H] < −5.0. Accordingly, we could stake the claim that this star was the first one to have undoubtedly formed from a gas cloud enriched by just one single first supernova. What a discovery—especially considering it was still the early days for the SkyMapper Telescope and its survey work. The results of SM 0313–6708 were published in *Nature* by Stefan Keller and colleagues in early 2014. Since then, the new star has already been providing inspiration and constraints for many new projects. We are now eagerly continuing the search for more stars with [Fe/H] < −5.0 and even [Fe/H] < −7.0 —stay tuned.

For completeness, it should be noted that several stars with about 1/70,000th of the solar iron abundance have also been found since the discovery for HE 1327–2326 in 2003. Such stars with [Fe/H] < −4.0 are out there, and with persistence, hard work, and a little luck we are discovering them one by one so we can use them to reconstruct the early Universe.

10.5 The Astronomical Community

The scientific field of astronomy is quite small compared to other sciences, such as chemistry or physics, for example. Consequently, it is no surprise that astronomers collaborate relatively closely with national and international colleagues on joint projects. The result is a very close-knit international community in which everyone knows almost everyone else.

There are many customs and conventions in this scientific landscape. One of them is inviting each other to deliver talks. The purpose is to inform one another about the latest research in all areas of astronomy and to promote scientific exchanges. I travel to other universities and institutions several times a year within the United States as well as abroad to give my presentations.

These invitations make it possible to personally report scientific results and to discuss them one-on-one with fellow members of the profession. One usually gets to meet many other astronomers on these trip which in turn leads to stimulating and interesting conversations, and new ideas and projects. Following my discovery of HE 1327–2326, I was regularly invited to give talks at colloquia and conferences. Over the years, I have given more than 100 scientific talks—ranging from popular science presentations to colloquium talks and plenary addresses.

At conferences, astronomers are sometimes in a mood to party. The floor shakes, the music booms—on a regular Wednesday evening. The entire bar is jam-packed with 300 dancing astronomers, from students to professors. A truly astronomical party is held on the second or third Wednesday in January every year. The occasion is the annual winter meeting of the American Astronomical Society (AAS), which up to 3,000 American and international astronomers attend to present current findings and consult about new results, projects, and initiatives.

The AAS has more than 7,000 members, who convene twice annually, in January and June. These meetings last four days and are among the busiest conferences I have ever attended. I am on my feet from early in the morning to late in the evening to listen to talks, meet with colleagues, and discuss projects or attend workshops. As these annual meetings are not confined to any scientific specialty, the participants represent every conceivable subfield in astronomy. This allows colleagues and friends to meet even when they do not work in exactly the same areas. As years go by you get to know many other astronomers, especially if you have been studying or working at different universities or observatories. For many of us these conferences resemble an "astronomical family" reunion.

At the winter meeting of 2011 in Seattle, I gave the Annie J. Cannon Prize Lecture to a packed house. I also gave a few interviews, shook many hands, and organized the "My GMT" photo drive at the booth of the 25-m large Giant Magellan Telescope in the huge exhibition hall. Details about this and other planned telescopes of the next generation are provided in chapter 11. Interested astronomers could pose in front of the large background image of the telescope and immediately take away a print of their portraits as a souvenir. We handed out some 200 photos and talked with many more people about this planned extremely large telescope.

These major conferences as well as smaller, more specialized meetings and workshops are extremely important and informative in spreading the word about the latest findings in science and also in exchanging ideas with international fellow professionals. They take place on all continents. Hence, astronomers travel a lot. If you are an observational astronomer, add to that observing runs at remote locations.

Astronomy clearly is among the most international of the sciences. Altogether, these aspects do make the astronomer's profession quite exciting and varied.

You also have to stay up to date daily about new findings by other scientists. A preprint server (http://arxiv.org/archive/astro-ph), also known simply as "astro-ph," is helpful in this regard. Most astronomers upload their scientific articles to this website. Some of these articles have already been accepted for publication in astronomical journals, others have just been submitted, and others are conference proceedings contributions.

All articles on this preprint server are available to the public free of charge, allowing rapid access to the latest findings.[1] A daily email notification informs about all the latest articles. The average lies at about 60 new articles per day from all subfields of astronomy. It is a daily ritual of many astronomers to check the astro-ph email for relevant and interesting articles. Sometimes new articles become the topic of big debates during coffee breaks at work the next day. Some departments even have regular astro-ph coffee discussions to talk about everyone's new publications. Facebook and Twitter are also widely used these days to share in real time the latest results and debate them with astronomers and laypeople alike. This system permits an enormously rapid, free dissemination of new results and findings around the globe.

---

[1] All of the author's scientific articles can also be found there.

CHAPTER 11

AT THE END OF A COSMIC JOURNEY

When studying the Universe, astronomers primarily rely on their observational findings. But not all puzzles can be solved by means of astronomical data alone. Theoretical calculations and computer simulations describing the physical and chemical processes of the Universe provide important supplemental information. Observational results from various research fields, when carefully combined with theoretical explanations, lead to new and more complete knowledge.

It is thus an important task to consider comprehensive simulations of how structure formed in the Universe, how galaxies evolved, and how these results tie in with the findings provided by metal-poor stars. Only then can the formation of our Milky Way's halo and the true origin of the most metal-poor stars be fully understood.

11.1 Cosmological Simulations

Our Universe is composed of 23% dark matter and 72% of dark energy. The luminous matter comprising gas, stars, and galaxies makes up a puny 5%. The exact nature of dark matter and dark energy is still not known, and the 5% luminous material is not easily understood either. But should we let ourselves be discouraged from studying the Universe just because the work of astronomers seems to divulge more questions than it delivers answers?

The answer is, of course, no. After all, we can use the knowledge we already have to inform the next steps. Dark matter, for instance, makes itself noticeable through its gravitation. Large masses—whether luminous or dark matter—always cause large gravitational forces. The effects of these forces can be observed and measured, even if the responsible mass itself cannot be seen. For this reason, astronomers often assume that galaxies are exclusively composed of dark matter. This is physically very simple to describe, because you can just work out the gravitation while not having to worry about the complex interplay between gas, star formation, supernova explosions, and the chemical evolution in each galaxy.

For this reason, large-scale simulations that were developed more than 10 years ago focused exclusively on the evolution of dark matter in the Universe from shortly after the Big Bang to the present day. This method resembles putting on 3-D glasses at the cinema so as to watch a movie in full 3-D. These simulations allow us to see what we could watch with a pair of "dark-matter glasses."

In such a simulation, the evolution of so-called dark halos is closely followed over time. These halos are regions where the dark matter is particularly dense. It is then assumed that a luminous galaxy together with its stellar halo would reside at the center of such a dark halo. This assumption is based on observations and the results of rotation curve analysis. They indicate that the Milky Way and all other galaxies are generally surrounded by a large halo of dark matter. Consequently, when simulating the structure and evolution of a dark halo, the evolution of a luminous galaxy can be reconstructed indirectly as well.

It should be noted that the term "halo" refers to a galaxy's thinly populated outer stellar region, whereas the term "dark halo" refers to an extremely large and dense volume of dark matter. Hence, a dark halo has nothing to do with the stellar halo of the Milky Way or that of any other galaxy. "Halo" merely designates an extensive spherically shaped object made of matter.

For the sake of simplicity, the formation and evolution of galaxies over the course of billions of years are still mostly simulated through the evolution of their dark matter halos. It would, of course, be much more instructive if such simulations would not only model the dark matter but also luminous matter. Indeed, the first such simulations are now being carried out, but all the processes involving luminous baryonic matter immediately become extremely complex. The running times of such simulations then become prohibitively long, exceeding the capacity of even the fastest supercomputers. These simulations can only approximate all those countless physical and chemical processes in galaxy formation. One alternative is to simulate specific processes independent from the overall cosmological evolution of the Universe, or else to conduct simulations that are temporally very limited, such as the formation of the first stars.

Cosmological simulations show in great detail how the first denser regions, the first dark halos, formed shortly after the Big Bang. Soon thereafter, some of these first halos merged into a somewhat larger halo assumed to have accommodated the first galaxy. Again, these and other smaller halos merged to form larger ones. Matter condensed into giant sheets and filaments, with halos forming at their intersections. Over time, some of these halos would continue to grow both by smoothly accreting matter and by gobbling up other halos to eventually become large galaxies, such as the Milky Way. Figure 11.1 depicts this large-scale structure of the Universe in a dark matter simulation.

Large-scale cosmological simulations detail the transition from a nearly structureless Universe after the Big Bang into a webbed Universe made from extended filaments over billions of years. These simulations reproduce various galaxy observations that have shown that luminous matter is not distributed evenly throughout the Universe. Instead, immense galaxy clusters coalesce from the web-like structures.

This long-lasting hierarchical assembly is the reason why many galaxies are still in a state of growth. Many observed galaxies show signs of past mergers and collisions. Such remnants can also be found within the Milky Way's halo due to its past mergers with smaller galaxies. The enormous gravitational pull of the Milky Way caused many galaxies, particularly smaller dwarf galaxies, to be captured and mercilessly torn apart upon entering its tidal force field. The remains of this consumption can be directly observed. Large photometric and spectroscopic surveys have discovered many huge, elongated streams of stars across the entire sky. They are remnants of disintegrated dwarf galaxies, stretched out like chewing gum. These observations proved that galaxies similar to the Milky Way nourish themselves and grow in quite a cannibalistic process of devouring smaller systems.

Not all smaller galaxies in the vicinity of the Milky Way suffered such a tragic end, however. The Local Group, with its many different dwarf galaxies, is a "side product" of the formation of our Galaxy. Most of

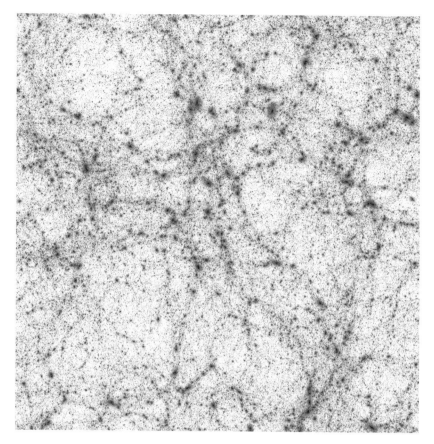

Figure 11.1. One snapshot from a cosmological dark matter simulation of the evolution of structure and galaxies. The dark matter manifests itself as a detailed filament structure within a region measuring 490 million light-years. The darker fields indicate dense crossing points of dark matter in which large galaxies and galaxy clusters are located. (*Source*: Printed with the kind permission of Brendan Griffen)

these surviving dwarf galaxies are orbiting either the Milky Way or Andromeda and will probably continue to do so peacefully for a long time to come. For others, their end is imminent. Some of the faintest galaxies show signs of distortion by the tidal forces of the Milky Way. They will be the next ones to be torn apart and their stars and gas will find their demise in the Milky Way's stellar halo.

From numerous observations it has long been known that any larger galaxy is usually being orbited by a number of smaller dwarf galaxies.

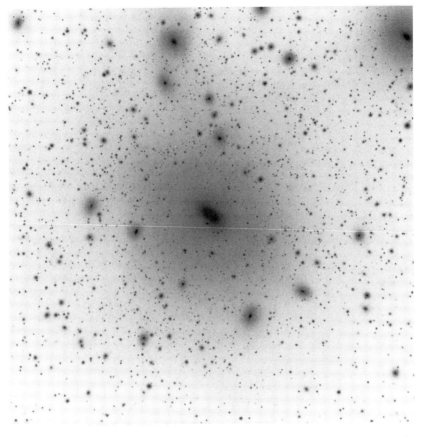

Figure 11.2. Snapshot image of a region 4 million light-years across in a cosmological dark matter simulation. At the present time, immense dark matter halos exist that host large galaxies such as the Milky Way. A multitude of smaller halos, assumed to be dwarf galaxies, are orbiting the central galaxy within its dark matter halo. (*Source*: Printed with the kind permission of Brendan Griffen)

In principle, this behavior is also visible in the simulations. Figure 11.2 shows a simulated present-day halo of dark matter large enough to host a Milky Way. It is surrounded by numerous smaller halos busily swarming around the central halo as if it were a giant beehive. Since each dwarf galaxy has its own dark matter halo, it can be assumed that the small simulated dark halos are the counterparts to the present-day observed dwarf galaxies. As suggested by simulations, one indeed ought to expect that large galaxies are orbited by many of dwarf galaxies. Thus, we can learn not only about the evolution of the Milky Way but also about dwarf galaxies and the interplay with their host galaxy.

These simulation results pose an interesting problem, though. Our Milky Way is surrounded by only about 40 dwarf galaxies, not over a hundred as found in the simulations. This discrepancy has been plaguing astronomers for over a decade. Many different solutions have been suggested, without success. The difficulty lies in the following: To what extent do these small dark matter halos host luminous dwarf galaxies that we can observe? Put another way, can it be possible that the Milky Way is actually surrounded by countless small dark matter halos that we cannot detect with our conventional telescopes? Or are dark matter simulations simply not detailed enough to describe luminous galaxies and their complex evolution?

With more sophisticated simulations, astronomers are trying to find answers to these fundamental questions. Observations can also contribute to understanding further details of galaxy evolution. New results on the nature of the various classical and ultra-faint dwarf galaxies help to reconstruct why some of them have survived the Milky Way's extremely cannibalistic growth while others have not. Finally, detailed observations of different stellar populations in the Milky Way provide crucial clues about individual steps in the long-term evolutionary history of our Galaxy.

The origin and evolution of galaxies have been fascinating astronomers for over half a century. Fundamental ideas about the formation of galaxies were developed starting around 1960, mainly focusing on the Milky Way. They were based solely on observations of stars and their motions within the Galaxy—long before the era of major cosmological simulations.

Two competing theories were developed. In the first model, a galaxy forms within a period of just 100 million years in a violently collapsing giant gas cloud. With only a few assumptions, the contemporary observations of element abundances in stars and their motions inside the Galaxy could be explained. The other model predicted that a galaxy would form only through mergers with small protogalaxies over a longer evolutionary period of several billions of years. Unlike the collapse model, this second model drew no connections between element abundances, star positions and motions in the Galaxy but predicted a completely hierarchical buildup of the galaxy. At about the same time as these two models came about, astronomers also found that every galaxy was embedded within its own halo of dark matter. This important fact, however, did not get incorporated into any of these early models.

As a consequence, two groups of astronomers emerged, each diligently gathering observations in support of their favorite galaxy model eager to disprove the other model. In particular, the motions of stars in the Milky Way's halo were examined to understand the processes that shaped the evolution of our home Galaxy.

In 1997, new observations led to the discovery of the dwarf galaxy Sagittarius, which reinvigorated the decades-long debate. What astronomers had found was not a bound and compact object, but a huge dense stream of stars extending across the sky that originated from a residual galaxy core. Hence, Sagittarius is not a whole galaxy anymore. It is currently completely disintegrating and, slowly but surely, being torn apart in the Milky Way's tidal force field.

The model of Galactic growth through mergers with other, smaller galaxies seemed to be confirmed by this new observational result. Today we know, however, that neither of these two models offers a complete explanation of the complex processes of galaxy formation such as that of the Milky Way. Cosmological simulations helped to show that an initial collapse phase occurs, which leads to the formation of the dark halos. They then continue to steadily grow by merging with other smaller halos.

Small dwarf galaxies are thus, in fact, direct witnesses of the Milky Way's formation and evolution. Their characteristics as well as their own evolution are marked by the dynamic interplay with their host galaxy— not only over the past 10 to 12 billion years, but also still today.

11.2 Where Do Metal-Poor Stars Come From?

This is one of the most exciting questions that stellar archaeology is attempting to answer. The prospect of employing metal-deficient stars in the Galactic halo to reconstruct the evolutionary history of the Milky Way has always motivated me a great deal. The oldest, 13-billion-year-old metal-poor stars must have formed before the Milky Way was what it is today. Most likely these ancient stars formed sometime during the earliest phases when the proto–Milky Way had just begun to evolve into a large spiral galaxy.

To gain insight into these early processes, it helps to again consider the detailed simulations of the Milky Way with its stellar halo and the hierarchical assembly of galaxies. Various simulations have shown that the inner part of the stellar halo formed quite early on, out of rather large "building blocks." Those could have been dwarf galaxies with a mass similar to that of the Magellanic Clouds. Due to their relatively large size, the gravitational tides are significant, causing such larger dwarfs to fall rapidly into the inner part of the newly forming host galaxy.

Today, the mean metallicity of the inner part of the halo is about one-tenth that of the Sun, hence $[Fe/H] \sim -1.0$. Qualitatively, this agrees well with the metallicity of the Magellanic Clouds. Although similar metallicities do not provide definitive support that such early merger processes had occurred in exactly in the described way, they nevertheless demonstrate that detailed knowledge about the dwarf galaxy properties is ultimately very important for our understanding of how the Milky Way formed.

Simply assuming that the more metal-poor stars are born before the more metal-rich stars does not actually answer the question of where the extremely metal-poor stars in the halo came from. However, it seems natural that at least some of these stars came from those early dwarf galaxies that met their ends in the inner part of the Milky Way. It is very likely that such building-block galaxies contained extremely metal-poor stars from their own, earlier evolutionary stages. After they fell into the Milky Way, most of their stars were then "unloaded." Hence, significant numbers of metal-poor stars are predicted to exist in the inner part of the Galaxy. But today, there are huge numbers of younger, metal-rich stars in the bulge and central region making the systematic search for the most metal-poor stars rather tedious—this is why astronomers have been concentrating on the halo. Nevertheless, huge efforts have finally yielded the first extremely metal-poor stars also in the bulge. Although extremely interesting, understanding the origin of metal-poor stars in the inner part of the Galaxy still does not help us understand where the metal-poor stars from the halo have ultimately come from.

We thus continue to look more closely at the evolutionary history of the Milky Way. According to cosmological simulations, the halo was repeatedly churned and heated by collisions whenever additional smaller dwarf galaxies were incorporated. These dwarfs were not massive enough to fall deep into the Milky Way's center. Instead the Galactic halo was filled with stars and gas from the torn apart and disintegrated dwarf galaxies—the Sagittarius dwarf galaxy is currently being shredded in the stellar halo in the same way.

Taking into account the hierarchical assembly of the Milky Way, it becomes important to more precisely examine the surviving dwarfs still orbiting the Milky Way at the present time. In particular, the faintest among them possess lower overall metallicities and have a higher occurrence of extremely metal-poor stars than the brighter, more massive dwarf galaxies. This behavior is described further in chapter 6. Under the premise that any devoured galaxies do not differ from surviving ones by some special trait, one would expect the ultra-faint dwarf galaxies to be particularly good providers of the halo's most metal-poor stars. We can even go as far as to postulate that all the halo's most metal-poor stars originally stemmed from these kinds of faint dwarf galaxies.

This hypothesis can be tested with observations of extremely metal-poor stars found in various dwarf galaxies. If all of the most metal-poor halo stars had been deposited into the halo by dwarf galaxies a long time ago, then they should have exactly the same chemical composition as equivalent dwarf galaxy metal-poor stars. If a significant difference in the element abundances is revealed, however, then perhaps either something is wrong with the largely successful simulations of structure and galaxy formation or the dwarf galaxies have survived for some, as yet unknown, reason.

My colleagues and I examined this important issue a few years ago. The goal had been to observe extremely metal-poor stars in some of the faintest galaxies known for the very first time with high-resolution spectroscopy to determine their detailed chemical abundances. Indeed, we could demonstrate that the element abundances of these stars and their counterparts in the halo are remarkably similar to each other. Figure 11.3 compares some element abundance ratios of metal-poor stars

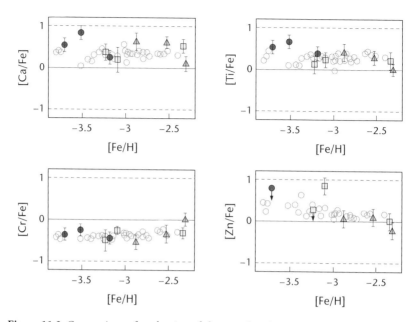

Figure 11.3. Comparison of a selection of element abundance ratios between metal-poor stars in the Milky Way's halo (white circles) and the least luminous dwarf galaxies. Squares: Stars in Ursa Major II; triangles: in Coma Berenices; shaded circles: in Segue 1, Boötes I, and Leo IV. The abundances of all these stars are remarkably similar to one another. (*Source*: Peter Palm; data from Frebel et al., *Astrophysical Journal* 708 (2010): 560–583)

in dwarf galaxies with those of metal-poor halo stars. Exactly as anticipated, the two groups of stars thus appear like "separated twins." This chemical resemblance can be viewed as a good indication for the oldest part of the halo of the Milky Way having been made, at least in part, from stars of former dwarf galaxies.

Figure 11.4 sketches how the buildup of the Milky Way's stellar halo could have happened and how the most metal-poor dwarf galaxy stars may have donated their stars to the halo.

With only a handful of extremely metal-poor dwarf galaxy stars available, it had been possible for us to provide observational evidence that supported the evolutionary events that drove Galactic halo formation long predicted by simulations. Furthermore, the individual element abundances showed that the chemical evolution had been inhomogeneous in these small, primitive, metal-poor galaxies. Since some of these

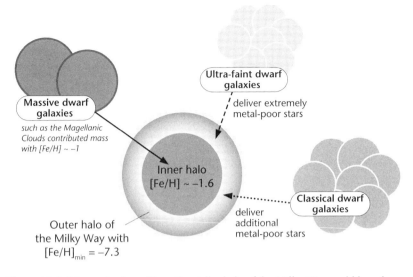

Figure 11.4. Schematic view of how the stellar halo of the Milky Way could have been built up. The most metal-poor stars we now observe in the halo probably came from various kinds of dwarf galaxies consumed by our Galaxy over time. (*Source*: Peter Palm, Anna Frebel)

stars have metallicities of [Fe/H] < −3.5, it is entirely possible that they formed from gas that had been enriched solely by some of the massive Population III stars. In that case, they would bear the chemical fingerprints of the first stars—after all, a first generation of stars must have existed in every galaxy.

Detailed cosmological simulations of the formation of the first galaxies find that Population III stars formed before and during the assembly of the first galaxies. This means that likely at least some of these first stars exploded inside the first galaxies. Hence, these first galaxies were probably not too dissimilar from the surviving ultra-faint dwarf galaxies. Then, chemical evolution operated for a relatively short period only, and further star formation ceased likely due to a lack of additional gas. This idea is supported by the fact that ultra-faint dwarf galaxies do not at all possess any metal-rich stars with [Fe/H] > −1.0.

Only improved simulations and additional observations will make it possible to unravel whether and to what extent a connection exists between the first and the surviving galaxies. The thought that ultra-faint galaxies could be surviving from the early Universe is extremely fascinating!

Despite these first successes, it remains very difficult to determine the chemical abundances of individual stars in any dwarf galaxy because of their great distances. Located deep inside the Galactic halo, their stars are extremely faint, making spectroscopic analysis very challenging. It takes a whole night of observing at a large telescope, such as the 6.5-m Magellan Telescope in Chile, to obtain a high-resolution spectrum of one star of acceptable quality.

But the promise of exciting results justifies the investment. By now, we have found over a dozen metal-poor stars in six different ultra-faint dwarfs. Each of these stars helps us to gain detailed knowledge about the overall stellar content of these systems, which, in turn, helps improve our understanding of the general evolution of dwarf galaxies.

To this day it remains unclear what drove the formation of the very first galaxies after the Big Bang and how the Milky Way began to form at that time. For the moment, it looks like chemical evolution always started off in the same fashion in every galaxy in the early Universe, or at least in a very similar way. It is exactly during this early phase in the evolution of a galaxy when the most metal-poor stars form. The individual size and mass of a galaxy may have defined how many stars would form over time, but nucleosynthesis and chemical enrichment seem to have proceeded independently of these other characteristics.

Yet, before concrete conclusions can be drawn about the most primitive generations of stars in dwarf galaxies, more of the most metal-poor stars need to be identified in as many different types of dwarf galaxies as possible. For a long time, extremely metal-poor stars have been sought in the somewhat brighter, classical dwarf galaxies known for decades. However, owing to inadequate search methods it had not been feasible to find these rare stars until a few years ago.

In 2010 my colleagues and I finally had a breakthrough. Thanks to new, sophisticated methods and several nights of telescope time, we were able to present the first extremely metal-poor star in the luminous dwarf galaxy that is located in the Sculptor constellation. Details about this discovery are described in chapter 7. Discoveries of additional such stars followed quickly. They all proved the existence of chemically primitive, metal-poor stars not only in the very faintest but also in classical dwarf galaxies.

Future observations of stars in to-be-discovered dwarf galaxies will either confirm the findings about the dwarf galaxy origin hypothesis of the most metal-poor halo stars or deliver interesting counter-suggestions. The stars identified so far suggest that they may well be members of the second generation of stars in each of their galaxies. The discovery and analysis of exactly these early stars is, after all, the goal of stellar archaeology.

In conclusion, the overall importance of the most metal-poor stars of the Milky Way to astronomy is steadily increasing. Due to their long lifetimes these low-mass stars are not only witnesses to the earliest processes of nucleosynthesis and the beginnings of chemical evolution, but also to the entire formation process of the Milky Way. Looking at them today, you would not think that they were the survivors of the many tumultuous events in their native galaxy and their violent incorporation into our Milky Way.

11.3 Expectations of Future Surveys

The most metal-poor stars are extremely rare. Sky surveys of the past decades have proven, nonetheless, that systematic searches combined with a number of selection steps lead to the successful identification of these objects. The subsequent selection of stars for follow-up observations depends on their magnitude, however. Faint stars require more telescope time or are generally too faint to ever be observed with high-resolution spectroscopy. For this reason, the outer part of the halo, beyond 15,000 to 20,000 light-years, is spectroscopically still largely unexplored.

Spectroscopic studies of the small dwarf galaxies orbiting the Milky Way have clearly shown that those galaxies contain extremely metal-poor stars, just as the halo. But they are located in the outer halo—the ones closest to us are some 130,000 light-years away. That means that only the very brightest of stars can be observed, namely the most luminous red giants. All of these brighter stars will be observed spectroscopically with the existing large telescopes in the coming years.

But what then? We need more data, particularly of dwarf galaxy stars, to study the chemical evolution of these tiny ancient systems and to investigate their importance for the evolution of the Milky Way. There are two potential approaches to solving this: either we need entirely new dwarf galaxies so that we can get more, sufficiently bright, just-observable stars, or else we simply need larger telescopes that can collect stellar light more quickly. Both these approaches are being pursued at the present time.

A large-scale sky survey is needed for a wide-ranging chemical characterization of the Galactic halo, including its various structures, streams, and dwarf galaxies. The Australian National University has initiated such a major project. The sky is currently being observed night after night, with the 1.3-m SkyMapper Telescope. It is located at Siding Spring Observatory together with the 2.3-m telescope and a few other telescopes and is shown in Plate 11.A.

This survey is completely automated, with the aim of digitally cataloguing one billion stars and galaxies of the entire Southern Hemisphere. The faintest objects to be observed will be a million times weaker than what we can see on the sky with the naked eye. New planets, stars, supernovae, galaxies, quasars, and other inhabitants of the cosmos will be discovered in the process. Every clear night, SkyMapper is recording 100 megabytes of data every 100 seconds. This means that, after five years, the entire Southern Hemisphere sky will be stored in 500 terabytes or 100,000 DVDs. These data will later become publicly accessible to everyone over the Internet.

Such a survey proceeds similarly to the photographic cartography of Earth. Separate images are taken of smaller regions of the sky, then merged and analyzed. Although research of metal-poor stars ultimately requires spectroscopic data, some SkyMapper survey data are already used for an initial selection. Unlike all previous photometric surveys, SkyMapper is specifically designed to enable the determination of the stellar parameters for each star. As SkyMapper is observing the sky with very specific filters for example in the blue, green, and red wavelength ranges, a sophisticated combination of the resulting brightnesses can yield information about an object's metallicity. The brightness of the star in the spectral region surrounding the strong, metallicity-dependent calcium K line near 3,933 Å is particularly important for these measurements. A strong calcium K line suggests a fainter star than one with a weak line due to less absorption of light. The fact that metal-poor stars appear bluer than metal-richer stars is also exploited. As a convenient result, metal-poor stars are shining brighter in this filter than most other stars. Such a distinguishing trait helps enormously when looking for that needle in the haystack.

Metal-poor candidates can thus be efficiently identified already from survey data, although the other two steps of spectroscopic follow-up observations are still necessary. Prospects are huge that the SkyMapper data will deliver many new results on stellar archaeology in the coming years.

Finding metal-poor halo stars is just one aspect that the SkyMapper telescope will facilitate. Discoveries of new star clusters, stellar streams, and even dwarf galaxies are also anticipated. Those new dwarfs will hopefully contain many brighter stars that would be observable with current telescopes.

Other sky surveys will be discovering new dwarf galaxies as well. The Dark Energy Survey (DES) is observing large portions of the Southern Hemisphere sky and has already delivered new candidate ultra-faint dwarfs. In Hawaii, Pan-STARRS, the Panoramic Survey Telescope & Rapid Response System, is currently in the process of mapping the sky with the first of two 1.4-billion-pixel cameras in search of asteroids near Earth. For comparison, my own small digital pocket camera has a lousy 5 million pixels. But whoever wants more pixels has to pay for it: astronomical cameras do cost astronomical sums. From 2023 the Large Synoptic Survey Telescope (LSST) in the Chilean Andes will then be repeatedly imaging the sky with its new 8.4-m telescope and 3-billion-pixel digital camera.

The designs of these surveys are less discriminatory than what the SkyMapper telescope is producing, however. Hence, they are not directly usable for a detailed characterization of stars. Although, new dwarf galaxies will be found, the metallicity information about the individual member stars will be insufficient. To determine metallicities, additional imaging will be required of all those stars—or better yet, spectroscopic observations.

Complementing these photometric surveys are other spectroscopic surveys. The Chinese LAMOST (Large Sky Area Multi-Object Fibre Spectroscopic Telescope) survey has been taking low-resolution spectra of objects in the Northern Hemisphere since 2010 from Xinglong Station situated 300 km north of Beijing. Metal-poor stars are currently being identified. The European satellite Gaia, developed by a huge team of astronomers throughout Europe under the direction of the European Space Agency, has been mapping the positions, distances, and velocities of one billion stars since 2013. Determinations of the stellar parameters and chemical compositions will be possible for a good part of all these objects. This information will also be important in the search for metal-poor stars and in characterizing the Milky Way, including its origin, evolution, structure, and dynamics.

But all in all, these new surveys are bound to produce a fire hose of new data that will enormously boost the field of stellar archaeology. Many of the stars will be too faint, though, for high-resolution spectroscopy with the current largest telescopes. We are already familiar with this problem. So we need to focus on the brighter stars in the halo and the new dwarf galaxies. Hopefully more stars with record-low iron abundances will soon be uncovered so that we can learn even more about the early Universe and the formation of the elements.

11.4 The Next Generation of Giant Telescopes

The fact that ongoing and new surveys are identifying many extremely faint objects in the halo and in various dwarf galaxies poses a great challenge for astronomy. At the same time, though, this problem prompts the strong desire to overcome these limitations and to look further out into the cosmos than ever before.

The need for high-resolution spectroscopy to analyze these as of yet unobservable, too faint stars might be alleviated in the next 10 to 20 years with any of the upcoming generation of giant optical telescopes. They will have mirror diameters of 25 m and larger and, consequently, will be excellently suited for the observation of interesting objects presently completely inaccessible to us.

Three such telescope giants are currently being built: a European one and two American ones. The European Extremely Large Telescope (E-ELT) will have a total diameter of 39 m, which will be achieved by interconnecting almost a thousand individual hexagonal 1.4-m mirrors. The completed mirror will look like a giant honeycomb. This project is directed by the European Southern Observatory (ESO), which also operates many other telescopes in Chile. The E-ELT will be built on the 3,000-m Cerro Armazones in the middle of the Atacama Desert in northern Chile. This site is 130 km south of the city Antofagasta and is just about 20 km away from Cerro Paranal, the site of the ESO's Very Large Telescope.

The mirror of the Thirty Meter Telescope (TMT) will be composed of 492 segments that can be fitted together to form a 30-m mirror. Universities in California together with partners from Canada, Japan, China, and India are operating this telescope. Next to the Subaru and Keck telescopes, it will be towering on the 4,000-m Mauna Kea on the Big Island of Hawai'i.

The Giant Magellan Telescope (GMT) will possess a total mirror diameter of 25 m. Its design differs from that of the other telescopes in that it will not consist of many small composite mirrors but of seven large, circular, 8.4-m mirrors arranged in a honeycomb structure. One mirror is at the center, with the other six arranged around it. The individual mirrors are as large as the largest mirrors today. Plate 11.B shows what this giant telescope will look like. It will be located at Las Campanas Observatory in Chile, which is currently home to the two Magellan Telescopes. The summit of the 2,500-m Cerro Las Campanas has already been leveled to create a large plateau for the construction of this telescope, which is led by a consortium that includes the American Carnegie Institution for Science and other universities and partners from across the United States, as well as Australia, South Korea, and Brazil.

Designing and constructing these new, exciting telescopes are expensive endeavors, with each telescope costing more than a US$1 billion. The operation of such a facility for 10 years costs almost as much again. This is why all these projects are major international collaborations involving many partner institutions to ensure that one day starlight will indeed fall onto those giant mirrors.

The TMT and the GMT are both supposed to be completed in the early 2020s, with the E-ELT following a few years later. They will be equipped with various instruments. Besides monstrous digital cameras, there will also be a high-resolution optical spectrograph—at least on the GMT, and later perhaps also on the other two telescopes. As the chair of the science working group composed of an international team of some 20 scientists, I have been involved in the planning phase of this GMT spectrograph. Our task was to prepare a detailed plan of the most innovative and promising scientific projects, ranging from the search for Earth-like planets to the most metal-poor stars to high redshift gas clouds. The required instrument specifications were discussed directly with the design team and then applied to make those envisioned scientific projects feasible. We enthusiastically developed plans for how to use the new spectrograph to find answers to hitherto unsolved scientific problems. It was simply thrilling to imagine how many new discoveries might soon become possible.

With this instrument, spectroscopists like myself will be able to look very far into the halo and determine the chemical composition of the most metal-poor stars in the outer halo. We will observe more stars in small dwarf galaxies and even approach more distant galaxies in the Local Group. We will also examine individual stars in both the Magellan Clouds for their composition and document the complex evolutionary history of these two galaxies.

We will obtain extremely high-quality data when observing any brighter stars with such a giant telescope. This could lead to magnificent new findings in nuclear astrophysics. Excellent data with a very high signal-to-noise ratio are required to make the smallest spectral details visible. Stars containing uranium could then be observed long enough for us to determine their ages from their spectra. Today this works only with very bright stars. All these observations would vastly broaden our horizon, in the truest sense of the word. They would tell us something about the history of early chemical evolution, and hence enable unexpected insights into the evolutionary history of the different kinds of galaxies.

All these future observations will then hopefully be interpreted in detail within the context of an improved theoretical understanding of the first stars and galaxies, of supernovae and element synthesis, gas-mixing processes, and star formation. New generations of sophisticated computer simulations which will be run on extremely fast supercomputers, will allow direct investigations of chemical evolution and the participating physical and dynamic processes of stellar systems, such as the very first galaxies. Such complex simulations will help discover whether, or to what extent, the faintest dwarf galaxies are related to the very first galaxies and whether galaxies such as these survivors are in fact the original "building blocks" of the Galactic halo.

11.5 Little Diamonds in the Sky

We have reached the end of our cosmic journey. We humans have only short lives compared to how long the Universe has existed. Nevertheless, we are part of it and descendants of the Big Bang and the stars. Our cosmic genes are the atoms that outer space has generated. Yet the beauty and elegance of the Universe and our ability to recognize it is more than just the material sum of atoms constituting everything. After learning about the many scientific details we can begin to reconstruct the fascinating evolution of the cosmos and are even able to clearly grasp all these grandiose processes.

In this quest, the oldest stars are patient companions. These individual, still surviving witnesses help us in a unique way to reconstruct how the very first cosmic events occurred. They do so on a very small scale, in terms of nucleosynthesis, as well as on a very large scale, pertaining to star and galaxy formation. As participants in these processes, the ancient messengers reveal to us how the very first generations of stars lit up the cosmos, how they died as gigantic supernovae, and how they set in motion the chemical evolution of the Universe.

The most exciting questions for which metal-poor stars deliver answers is the complex interplay among the chemical, physical, and dynamic processes that have operated for billions of years to shape the evolution of our Milky Way and the Solar System including our planet

Earth. This also yields insights into the evolution of elements such as carbon. We can thus link the chemical evolution of the cosmos with the biological evolution on Earth—both are indispensable to the emergence of life.

The many results based on the work with metal-poor stars continue to advance other subfields of astronomy as well. At the same time, though, many unsolved problems remain, which astronomers aim to answer in the years to come, particularly with the help of major sky surveys, new gigantic telescopes, and enormous computer simulations. The wonderful thing about working with metal-poor stars is, and will remain, that there are so many different ways of gaining new exciting insights into the history of Universe and our Milky Way. For each problem there always seems to be a suitable metal-poor star.

At the end of many of my presentations I like to refer to metal-poor stars as the little diamonds in the sky: you very rarely find one, many of them contain large amounts of carbon, they sparkle for billions of years, and if you find one of your very own, you can count yourself lucky and fortunate. In closing, let me just give a variation on Marilyn Monroe: "Metal-poor stars are a girl's best friend!" Who still needs diamonds?!

Further Reading ✦ ✦ ✦

Popular Science

Brainard, Curtis. "The Archaeology of the Stars." *New York Times*, February 10, 2014. http://www.nytimes.com/2014/02/11/science/space/the-archaeology-of-the-stars.html?_r=0.

Croswell, Ken. *The Alchemy of the Heavens: Searching for Meaning in the Milky Way.* New York: Anchor, 1996.

Frebel, Anna. "Is the Milky Way a Cannibal? An Astronomer Travels to the Driest Place on Earth to Find Out." *Scientific American* 307, no. 6 (2012). http://www.scientificamerican.com/articleis-the-milky-way-cannibal-one-astromer-travels-driest-place-earth-find-out/.

———. "Reconstructing the Cosmic Evolution of the Chemical Elements." *Daedalus* 143, no. 4 (2014). http://arxiv.org/abs/1408.4832.

In-Depth Reading

Frebel, Anna. "Stellar Archaeology: Exploring the Universe with Metal-Poor Stars." *Astronomische Nachrichten* 331, no. 5 (2010): 474–488. http://arxiv.org/abs/1006.2419.

Frebel, Anna, and John Norris. "Metal-Poor Stars and the Chemical Enrichment of the Universe." In *Planets, Stars and Stellar Systems*. New York: Springer, 2013. http://arxiv.org/abs/1102.1748.

———. "Near-Field Cosmology with Metal-Poor Stars." *Annual Review of Astronomy and Astrophysics* (forthcoming). http://arxiv.org/abs/1501.06921.

More about Stellar Archaeology and the Author's Work

All articles by the author can be found on the preprint server: http://arxiv.org/find/astro-ph/1/au:+frebel/0/1/0/all/0/1.

Author's webpage: http://www.annafrebel.com.

Search for "Anna Frebel" on the Internet and many articles about various research results can be found.

INDEX ✦ ✦ ✦

Pages in *italics* refer to figures.